T0332239

# LARGE SPACE STRUCTURES FORMED BY CENTRIFUGAL FORCES

**Earth Space Institute Book Series**
Editor in Chief: Dr Peter Kleber, Chairman, Earth Space Institute

The Earth Space Institute is a division of the Foundation for International Scientific and Education Co-operation, registered in London, no. 2254798. The Institute is a non-profit organization.

*Volume 1*
Space Science in China
edited by *Wen-Rui Hu*

*Volume 2*
Der Mensch im Kosmos
edited by *Peter R. Sahm* and *Gerhard Thiele*

*Volume 3*
Multiple Gravity Assist Interplanetary Trajectories
by *Alexei V. Labunsky, Oleg V. Papkov* and *Konstantin G. Sukhanov*

*Volume 4*
Large Space Structures Formed by Centrifugal Forces
by *Vitali M. Melnikov* and *Vladimir A. Koshelev*

This book is part of a series. The publisher will accept continuation orders which may be cancelled at any time and which provide for automatic billing and shipping of each title upon publication. Please write for details.

# LARGE SPACE STRUCTURES FORMED BY CENTRIFUGAL FORCES

VITALI M. MELNIKOV

VLADIMIR A. KOSHELEV

*Korolev Rocket and Space Corporation Energia*
*Moscow, Russia*

Translated by Nina Barabash

**CRC Press**
Taylor & Francis Group
Boca Raton London New York

CRC Press is an imprint of the
Taylor & Francis Group, an **informa** business

Cover illustration: The Znamya-2 experiment. The spacecraft with unfolded sail eight minutes after its separation from the Mir orbital station. The image was obtained from the orbital station at a distance of 160 m.

First published 1998 by Gordon and Breach Science Publisher

Published 2020 by CRC Press
Taylor & Francis Group
6000 Broken Sound Parkway NW, Suite 300
Boca Raton, FL 3487-2742

© 1998 by Taylor & Francis Group, LLC
CRC Press is an imprint of Taylor & Francis Group, an Informa business

No claim to original U.S. Government works

ISBN-13: 978-90-5699-112-8 (hbk)

**Visit the Taylor & Francis Web site at**
**http://www.taylorandfrancis.com**

**and the CRC Press Web site at**
**http://www.crcpress.com**

**British Library Cataloguing in Publication Data**

Melnikov, Vitali M.
   Large space structures formed by centrifugal forces. -
   (Earth Space Institute book series; v. 4)
   1. Large space structure (Astronautics) 2. Centrifugal force
   I. Title II. Koshelev, Vladimir A. III. Earth Space Institute 629. 4´7

ISSN: 1026-2660

# CONTENTS

# PREAMBLE

Of all the problems that face the whole Earth, global power engineering and transportation are becoming the most crucial. Energy is the key to industrial development, transportation, commerce, healthy economy, and high living standards. This book is giving approaches to encourage further search for feasible ways to solve the problems.

The global nature of these objectives requires non-standard approaches and pooling the efforts of the world community for successfully coping with the problems in the given area.

RSC Energia, a leading corporation for space technology in Russia, was founded by the eminent designer, Academician S.P. Korolev, in 1946. The name of S.P. Korolev is inseparably linked with the novel solutions in space technology paving the way for the first generation artificial Earth satellite and offering man the unique opportunity of breaking away from the Earth's gravitational clutches.

Throughout this period RSC Energia paid exceptional heed to the matter of space exploration for peaceful purposes. The last decade is prevailing in the development and utilisation of manned orbital stations serving for quite a long period as a platform for international co-operation.

In 1980 RSC Energia commenced works on the development of large space structures expanded by centrifugal forces. This trend owes much to the authors of this book in their leading role, ideological aspects, and design activities.

To successfully complete the important milestone – Znamya-2 experiment for deploying the film reflector of $D = 20$ m in space – it was necessary to combine efforts and experience of various divisions of RSC Energia to support the developmental works on the experiment within the 6-year period. The Design Divisions, Instrumentation Products Division of the Experimental Machine Building Plant, Testing Division, Cosmonauts Training Centre, Mission Control Centre, and Baikonur Launch Site were involved at different stages of the experiment development. It is merely due to the co-ordinated activities of all participants that this unique experiment has been implemented. Technical feasibility of such structures offers a new trend of advanced, high space technologies.

We are very pleased with the fact that Gordon and Breach Science Publishers, understanding the necessity of international co-operation in achieving these goals, set about publishing and distributing this book in many countries thereby attracting the attention of the world scientific community.

N.I. Zelenschikov,
The First vice-president,
The First Deputy Designer General

# CHAPTER I

# INTRODUCTION

This book reviews the comprehensive studies of problems dealing with the development of large space structures expanded by centrifugal forces. An important role of large space structures expected in future emerges from the outlook for the development of power engineering within the period of 2000—2050. The principal motivations behind this future developmental activities are:

• the 2—3 fold increase in the energy consumption against the current level, primarily, due to the development of the third-world countries,

• a decrease in the oil, gas, and coil consumption due to the depletion of the resource base and ecological consequences,

• a restricted use of nuclear power stations because of difficulties in the nuclear wastes utilisation,

• the accommodation of energy-consuming and ecologically hazardous industrial productions in the Low Earth Orbits and on the Moon surface,

• a widespread use of particularly inexhaustible power sources through the development of solar power stations, solar concentrators, and space-based reflectors both to support space-based industries and to transfer energy to the Earth.

Thus, the large, space-based, sunlight reflectors of several hectares in area would contribute greatly to the future power engineering and space technology. However, even now these structures could facilitate the solution of several important tasks such as:

• energy retransmission, TV and radio communications,

• illumination of the Earth regions with reflected sunlight (areas beyond the Polar Circle, sites of catastrophes, etc.),

• local weather and crop management,

• debris removal from space,

• development of solar sails for interplanetary missions,

• development of VLF and ELF antennas for ionospheric research,
• communications with deep-water vehicles,
• development of parabolic concentrators and radio antennas.

The ideas of solar sailing were first introduced and motivated in the twenties by the founding fathers of modern rocketry, F. Tsander and G. Oberth. However, only nowadays these ideas have become feasible.

Considering the problem of retransmitting, it should be noted that, with large data streams routed between a large number of users, certain advantages are offered by passive retransmitters (such as reflectors under consideration) against the space-based active retransmitters.

The arguments put forward in favour of illumination of regions beyond the Polar Circle from the space-based sources are:
• a psychological effect due to the illumination of the Earth regions by reflected sunlight during the polar night period,
• ecologically pure space-based illumination sources as compared to the artificial electric light sources of only 1% efficiency,
• cost saving, fuel delivery, or routing of electric lines within these areas are connected with large expenses and ecological violations,
• a wasteless power generation process.

In the years of space activities the mankind has brought so many artificial objects in the Earth orbits that the danger of a collision with space debris has become a serious threat. Proceeding from the current predictions, spaceflights could become impossible even by 2000 because of a high risk of collisions. This is the main motivation to make the removal of space debris a matter of vital question. However, such debris hitting a film screen at relatively low velocities would be crushed into fragments of safe sizes and transferred to other orbits.

The development of large space structures of several hectares in area and their attitude control is a complicated science-engineering task having no analogues in the ground and space technologies and requiring unique approaches to be successfully implemented.

By combining space conditions (deep vacuum, microgravity, solar fluxes) and principles of surface shaping by centrifugal forces new opportunities are offered for creating "truly" space-based large reflectors, antennas, and concentrators.

A concept of building a film surface (unfolding and keeping stiff) by centrifugal forces would save the need of using the supporting structures required to make the surface rigid. The important advantages of this concept over analogues using a supporting structure are:
• a low mass-to-surface area ratio, $\rho = 5-10 \text{ g/m}^2$,
• a small stowed volume during transportation and an automatic in-orbit deployment with a low energy consumption,
• a capability to form surfaces as large as $10^5 \text{ m}^2$ with a small-size deployment mechanism,
• a non-propulsive attitude control,
• a high accuracy of surface configuration,

• a simple and reliable structure and relatively low cost.

Fifteen to twenty years ago efforts were mainly focused on increasing a mass of payload to be placed into orbit. Now space hardware is imposed high performance requirements including minimum expenses for the development of new technologies required to cope with primary economical and science problems. Large space structures expanded by centrifugal forces meet these requirements.

The objective of this work is to make a comprehensive research in the technology enabling the development of space-based structures expanded by centrifugal forces, dependent on the nature of these forces, environment, and dimensions of the structure to be deployed.

To design and develop such application specific vehicles, efforts are required to:

• work out general concepts, design criteria, development steps,

• select sound structures of film surfaces, folding patterns providing a smooth (with no impacts, oscillations, tangling), controlled deployment from a folded pattern to a required configuration and controlled repackaging,

• provide the controllable deployment and repackaging dynamics through specially designed mechanisms, mathematical simulation of the deployment and repackaging processes, dynamic simulation during ground tests,

• assure attitude control of the deployed structures in compliance with the task specific requirements for the surface accuracy and orientation, passive and active damping of oscillations caused by gyroscopic forces, to select a combination of structure parameters allowing oscillations on a case-by-case basis,

• provide effective reflecting coatings and strength performance of film materials simultaneously subjected to operational loads and environmental exposure during a long operation period (deep vacuum, cyclic temperature from $T_{max}$~100°C, UV radiation, fluxes of protons, neutrons, etc.),

• develop technologies for manufacturing and packaging film materials (select patterns, modes of welding or bonding to produce seams equal in strength to the base material, to provide manufacturing accuracy, and minimise stowed envelope),

• select proper fastening schemes and winding patterns for tether elements of reflectors and tethered structures, as a whole (antennas, flywheels), to provide dynamics of their controllable deployment, and to avoid tangling,

• make systems and assemblies of the structure expanded by centrifugal forces compatible with conventional systems of the vehicle, to conduct their integrated ground tests.

Figure 1.1 shows stages of developing a vehicle containing transformable elements. Key issues encountered in designing such vehicles are:

• deploying from a stowed configuration and repackaging,

• re-orienting a deployed vehicle in space,

• assuring strength and reflection capabilities of structural materials,

• designing the deployment and repackaging mechanisms and transformable elements,

• ground and flight testing.

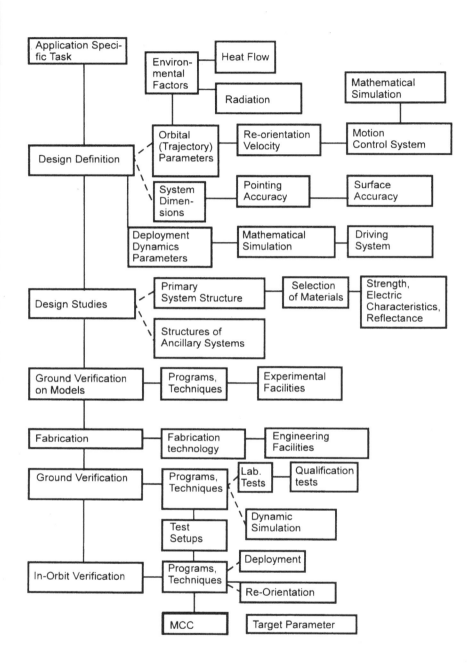

Fig. 1.1. Stages of developing a vehicle

With respect to the aforesaid application specific tasks, the structures expanded by centrifugal forces possess design features depending on the requirements dictated by the application specific task to be solved, surface accuracies, and orientation.

Structures of reflectors, concentrators, solar sails, and shields use both film (mesh) and tether elements. In tethered, framed antennas and flywheels only tether elements are employed.

A vehicle carrying a deployable structureless system, besides the convenient systems, is provided with a container in which a stowed structure, its deployment mechanism including its power supply and control equipment are delivered to orbit.

The reflector surface accuracy and attitude control requirements for retransmission and communications are not very stringent. For the UHF and VHF waves the surface may be of mesh material with a mesh size below $0.05 \lambda$ (where $\lambda$ — wavelength). Here, a wide radiation pattern would allow an orientation tolerance of several degrees.

On the contrary, the sunlight reflecting surfaces intended to illuminate the Earth regions are imposed extremely stringent requirements for the surface shape accuracy and orientation (below 10′) that is governed by a necessity to provide a maximum energy concentration on a light spot. A surface of a maximum reflecting capability would be required (e.g. sodium film periodically deposited in orbit on polyimide substrate which offers a reflection factor of 0.98).

A parabolic solar energy concentrator also requires accurately configured surface of a high reflection performance. On the contrary, the UHF and VHF parabolic antennas could be made of wide-meshed metal material imposed low requirements for a surface accuracy. The structure could be composed of several sectors, or cells joined so that clearances would be up to $0.05 \lambda$ .

In designing a solar sail no stringent requirements are imposed on a surface accuracy. The surface could be made up of separate sectors or strips, both coupled and non-coupled around the periphery. The sectors could be rotating about the radial axis thereby allowing one to control the structure through the centre of mass offset relative to a centre of sunlight pressure (Fig. 1.2).

The space debris isolation shield is imposed the lowest requirements for a surface accuracy due to the angular isotropy of the space debris velocity distribution.

Wavelengths of the ELF and LF loop antennas run to hundreds of kilometres, however, the ambient space plasma decreases the wavelength and, to properly set the antenna to a generator, the envelope area has to be changed. It is also required that different parts of the envelope be within one plane, most often within an orbital plane. During deployment and operation the tethered antennas can be of the loop or circular configuration (Fig. 1.3).

A flat-disc configuration is preferable for reflectors and shields. Parabolic antennas and concentrators are paraboloids of revolution (Fig. 1.4).

Projects of large systems such as a solar sailing vehicle for missions to Mars with the sail diameter of 200 m and sunlight reflectors for illumination of the Norilsk and Yamburg regions during the polar night have been developed. Besides, the tethered ELF antenna concept has been proposed.

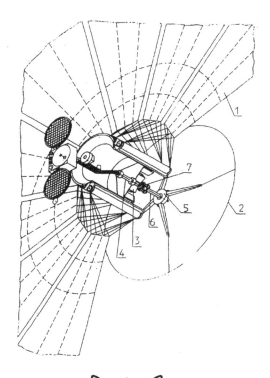

Fig. 1.2. Solar sailing vehicle. 1 — film surface built by centrifugal forces; 2 — tethered counter-rotation flywheel; 3, 5 — sail and flywheel rotation electric drives; 4, 6 — rotation axes; 7 — hinge of rotation axes deflection

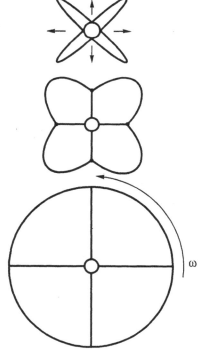

Fig. 1.3. Tether antenna of the loop and circular configuration

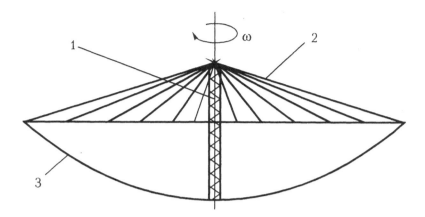

Fig. 1.4. Parabolic antenna (concentrator). 1 — expandable truss; 2 — guy-tethers; 3 — reflecting surface

To verify the computing techniques and concepts incorporated in the projects of large space structures expanded by centrifugal forces, the space-based Znamya-2 experiment has been developed and implemented on 02.04.93 aboard the Progress transportation and cargo vehicle where the 20 m reflector has been deployed (Fig. 1.5).

Fig. 1.5. Space-based reflector. (Photo taken from the long-term Space Station Mir)

The world mass media presented the news of the successful accomplishment of the Znamya-2 experiment as a new trend opened up in the area of space technology. The first-ever large reflector has been deployed and the unique experimental program has been implemented.

The success of the Znamya-2 experiment gave rise to a new endeavour and demonstrated the feasibility of large space systems.

# CHAPTER II

# ANALYTICAL OVERVIEW.
# STATUS OF DEVELOPMENTS

Large space structures can be classified the most completely stemming from their deployment principles. Based on this approach, such structures fall into the following groups:

• mechanical, where structural elements are hinged into one kinematic system; when stowed, the elements are closely packed and held together with special links; after the links are released, the system elements are expanded into a specified configuration by spring energy,

• rotating, where deployment and shape-keeping rely on centrifugal forces generated by rotating the structure,

• pneumatic, where deployment and shape-keeping are provided by a residual pressure inside the structure,

• specific, where a shape is maintained through creating an electrostatic field,

• combined, where the structure is deployed by pressure or spinning and the shape is maintained at the expense of a supporting structure (a frame).

Nowadays, the mechanically deployed structures are widely introduced. Their chief advantage is a specified configuration accuracy achieved through stiffening the structure.

A drawback of the mechanically deployed structures is a relatively low deployment coefficient (a ratio between the structure dimensions in stowed and deployed configurations) which limits their use in designing large objects.

Large deployment coefficients (of the order of ten and even hundreds) could be obtained by using inflatable structures.

A disadvantage of such structures is the probability of a micrometeoroid hit which would violate the structure geometry and lead to its failure.

The aforesaid drawbacks could be eliminated by deploying the structure into specified configurations through the use of compression members whose stiffness in a deployed configuration would be provided at the expense of tensile stresses generated in them under centrifugal forces.

Materials used to fabricate antenna reflecting surfaces have to meet specific requirements dictated by structural features of foldable antennas and by a necessity of operation in space environment. The requirements are: flexibility and elasticity, minimum mass, required strength, resistance to space environment.

Among the most advanced materials, the current composites based on graphite fibre, kevlar, or boron carbide with epoxy resin are considered. Also thought as candidate materials are metallized textiles, metallized polymeric knitted materials, metal fibre knitted materials.

The antennas designed as four tethers 39 m long with weights attached to their ends, which have been deployed on the US artificial earth satellites Explorer-4 and IR-2 at the expense of the satellite body rotation, must be considered the first (early sixties) large structures expanded by centrifugal forces.

The inflatable satellites-balloons have been also designed among the first, large, structureless systems deployed in space.

For example, the Echo-1 passive relay satellite designed by the Langley Research Centre (NASA) as an inflatable balloon $D = 30$ m was made of the Mylar-type polymer film 13 μm thick, covered with aluminium coating of 2200 Å. An initial pressure inside the coating was $2.8 \cdot 10^{6}$ atm, the surface temperature was varying from $-62°C$ to $+115°C$.

This satellite had been launched in 1960 and was staying in orbit of 1600 km for eight years despite the fact that it had been rated for a one-week operation. Such a long life period had resulted from the fact that, upon the complete loss of strength, the film material did not dissociate because of the lack of environmental effects (except for light pressure). However, the satellite has been undoubtedly penetrated by meteorites more than once.

Such satellites-balloons have been also employed for purpose of satellite geodesy.

In 1990 the Zond-2 experiment has been conducted aboard the Progress transportation and cargo vehicle using scientific hardware designed and manufactured by the Tashkent Design Office for Machine Building. The objective of the experiment was to inflate a balloon of $D = 10$ m.

However, because the deployment dynamics had not been properly verified, the balloon burst thereby causing the vehicle and the coating to get into a tangle. Eventually, the remainder of the film balloon was separated from the Progress vehicle.

One of the first large, framed structures deployed in space was the KPT-10 radiotelescope accommodated on the Salyut-7 space station transfer compartment hatch in 1979.

The radiotelescope contained a parabolic mirror of $D = 10$ m made of a metal screen stretched on a supporting structure (a frame) consisting of a large number of collapsible elements. The telescope has been deployed in space by means of compression members. This telescope enabled to receive signals within the range of $\lambda = 12$ cm and $\lambda = 72$ cm. Combined with the ground radiotelescopes it has been used in the capacity of an element of an interferometer with a large baseline. In the process of the KPT-10 jettison from the space station the telescope was entrapped by the station structure elements. As a consequence, the station crew had to perform extravehicular activities (EVAs) to recover from this contingency (the entrapped tether was severed).

Works were initiated on the expensive design of radiotelescope KPT-30 containing the antenna of D = 30 m. The KPT-30 telescope would be deployed by means of 12400 spring members. However, this design has not been implemented.

In 1986 solar arrays have been deployed by a spring mechanism on items 17 KC and 77 КСД, the deployment proceeding as predicted. In 1987, from the 17KC item airlock, the lentil-shaped truss 10 m long has been automatically deployed using an electric drive. In 1989 the multi-use solar array deployment system has been employed for the controlled deployment of two 15 m solar arrays using an electric drive.

A structure, similar to the last one, has been used in the US and French projects of solar sailing vehicles designed for missions to Mars. The projects have not been implemented, although the ground verification of framed film structures has been conducted.

In a number of experiments on the deployment of large truss structures in space it was assumed to involve cosmonauts. In 1987, during two EVAs, an additional solar array was installed by the crew on item 17KC N121. In 1990 the telescopic cargo boom was brought out and mounted on Space Station Mir. In 1991, on Space Station Mir, a truss structure of 12 m using the plastic recovery materials has been assembled and deployed (the Sofora experiment).

An experience gained from the ground tests and space-based automatic deployment of a number of circular shape antennas of up to 30 m in diameter on the Progress vehicle is of a particular interest. These are such experiments as "Model" using a backbone of elastic profiled material (1980); "Model-2" using a backbone with a conductive tape (1981). In 1983, under the same name, a space-based experiment has been performed using a modified backbone and a conductive tape. In all the three experiments the listed structures have not been successfully deployed.

In 1986 the experiment, again named "Model-2", for deploying a large inflatable antenna in orbit has been conducted. The antenna was designed as two rings of $D = 30$ m positioned symmetrically relative to the vehicle longitudinal axis. Eventually, the structure has been fully deployed, however, the deployment procedure departed from the predicted scenario. Early in the deployment the structure started to be deployed "in bursts", rather than smoothly in compliance with the predicted scenario. This created a danger for the vehicle to be entrapped by dozens of meters of the rubber sleeve. However, by chance, jamming or tangling had not occurred and, upon pressurization of the antenna sleeves, the structure was deployed and operated in the capacity of the LF transmitting antenna.

While preparing for the space-based Model-2 experiment, the ground deployment verification of the inflatable structure was conducted in a large neutral buoyancy facility (by hanging) in a spacious room, as well as in a vacuum chamber.

In the neutral buoyancy facility we succeeded in smoothly deploying the rubber-impregnated sleeve folded into the bellows-type pattern and stowed in a container. The sleeve was unfolded by air pressure supplied from inside. The deployment was controlled via two battens holding the entire package. The deployment velocity depended on how strongly the battens were holding the package to be unfolded.

The structure has been deployed in a large, covered sports hall and its deployed configuration was under control. It was hanged by means of a large number of attached meteorological balloons. The structure was configured into a regular, circular shape at a low degree of its out-of-plane. Rings of a near-plane circular shape were obtained during the actual space-based experiment that was in contradiction with theoretical predictions saying that the figure-of-eight configuration would be inevitable with structures of a small overall height, that is with a very small ratio between the diameter of the rubber impregnated sleeve, $d = 0.15$ m, and the diameter formed by the sleeve in space, $D = 30$ m. Here, $d/D = 0.005$, i. e. very small. To maintain a plane configuration the $d/D$ ratio of about 0.1 was assumed.

During tests in a vacuum chamber the deployment "in bursts" was observed when the structure was "jettisoned" from the container at a large velocity due to a residual pressure within the sleeves and insufficient spring-loaded forces of the structure holding battens. An extent of tightening was difficult to control and it was just the mechanism of deployment "in bursts" realised in the experiment. The sleeve outer layers, once the holding links had been released, expanded from under the holding battens at a large velocity carrying away the remainder of the structure. In a matter of fractions of a second the sleeve, folded in two, expanded at the full length normally to the vehicle longitudinal axis and then, in the opposite motion, stroke against the vehicle and got into a tangle. However, by the action of pressure supplied to the sleeve it started to separate from the vehicle in a sector-by-sector manner not getting into a tangle with the vehicle structure elements and took a regular circular configuration. The second symmetric circular part of the sleeve-antenna was similarly deployed on the other side of the vehicle thereafter.

It must be noted that the deployment step has not been properly verified despite the ground testing and simulation performed over a long period of time.

The Crab experiment was the last step in a set of verification tests of the Model-type structures. The experiment objective was the in-orbit deployment of the loop antenna at the expense of using plastic recovery materials.

Here, again an off-nominal deployment was observed. Because of the lack of power aboard the vehicle, the plastic recovery elements were not heated to the estimated level, however, after several orbits around the Earth the structure was additionally heated by the Sun and, eventually, deployed completely.

Also worthy of note are unsuccessful experiments on the deployment of truss antennas of 4 m in radius on small free-flying satellites, MAK-1 and MAK-2, placed into orbit from the Mir airlock in 1992.

Of a particular interest are works on a solar sailing vehicle model carried out in the USA in parallel with the development of the Znamya-2 experiment at RSC Energia. This model was designed as a framed structure containing 4 flexible struts expanded from a central cylinder serving as a transport container for a solid, rectangular sheet of a film reflector. The rectangular side was 20 m long (the Znamya-2 experiment reflector was also 20 m in diameter). It was assumed to deploy the structure aboard the Space Shuttle (Fig. 2.1).

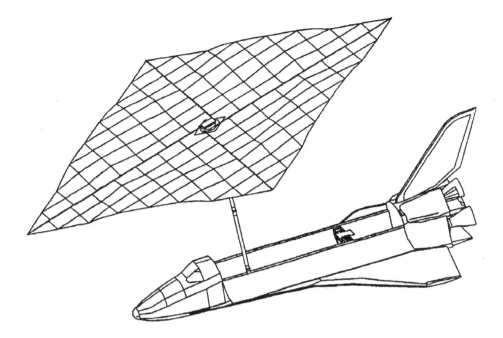

Fig. 2.1. Predicted deployment of the solar sail model on space shuttle

Thus, in 1990 a prototype solar sailing vehicle has been designed and subjected to ground tests in the USA. However, its assumed launch dates have not been announced thus far.

Due to the preceding accident and loss of astronauts, as well as high cost of space-based experiments conducted aboard the Space Shuttle, Lois D. Freedman, Executive Director of the American Planetary Community, in 1990 declared his intention to RSC Energia to investigate the possibility of testing the US-developed solar sail model aboard the Progress transportation and cargo vehicle. Having considered this proposal, RSC Energia gave a positive response, however the US side did not pursue the initiative.

The tethered systems can be considered as one of the option of large space-based structures. On August 5, 1992, in the Space Shuttle STS-49 mission, an attempt was made to aloft into orbit the $ 379 million-worth Italian research satellite to be towed with a tether of 19 km long. However, once the tether had been unreeled to 262 m, it was jammed in the reduction gear. Efforts had been made to proceed with the deployment, but failed, and the satellite was pulled back into a cargo bay.

In February 1996, an experiment came to grief when an attempt was made to deploy a tethered satellite into orbit from the Space Shuttle. Once the deployment had been completed, the 20 km tether broke away from the Space Shuttle together with the satellite.

The most large-scale project of the solar sailing vehicle heretofore is the "geliorotor" project [1] intended for the Halley rendezvous mission (1978). The vehicle was designed as a solar sail, expanded by centrifugal forces, made up of 12 strips like rotating helicopter blades, each 8 m wide and 7500 m long. The project suggested that the sail would be fabricated of aluminised Mylar film stretched on a light, transformable titanium backbone and deployed by the solar radiation pressure with the initial spinning from plasma thrusters. The $62 million project had not been implemented because of a high risk level.

The international competition for designs of solar sailing vehicles announced in honour of the 500th anniversary of America discovery by Columbus (1990) has summarised foreign designs based on non-rotating supporting structures which has not been implemented to date.

Analysis of theoretical works showed a good start in mechanics of transformable systems sufficient for the initiation of design activities.

It must be noted, as a whole, that the experimental works are not keeping pace with other areas of space technology, e. g. with manned or unmanned vehicles. The inadequate design studies and ground verification resulted in contingencies occuring in most of the in-orbit experiments. The design failure history was remaining within the reach of a few design offices. Such a situation is usual for the first steps into new trends. Large-scale projects have not been implemented hitherto.

CHAPTER III

# DYNAMICS OF DEPLOYMENT FROM STOWED CONFIGURATION AND REPACKAGING

## 3.1. Packaging and Deployment Considerations

*3.1.1. Folding Pattern, Rotational Moment, and Deployment Velocity Requirements*

In coping with the problem of deploying a flexible system (film or tether) from a stowed configuration by applying centrifugal forces, the following basic requirements should be met:
 • while being deployed, the structure should be insensitive to perturbations which could unrecoverably violate the shape,
 • minimised mass-dimensional properties of the stowed envelope,
 • packaging efficiency,
 • no closed volumes with entrapped air,
 • capabilities should be provided to remove air from any sector of the package during evacuation,
 • a proper method should be selected for a torque application (point of application, a value or a law of change) enabling the deployment with no intolerable oscillations, entanglement, and damage to the structure,
 • the equipment should be selected so as to assure the required torque and reliable ground verification,
 • a law of deployment velocity variation has to be established and equipment to support this law and to be verified under the ground conditions has to be identified,
 • additional equipment preventing from or eliminating the possible entanglement of a system (freewheel or momentum clutch, feedback) should be identified,

• reliable interfaces should be provided for the equipment assuring the experimental equipment deployment in parallel with other systems of a space vehicle.

Certain tasks impose limitations on the system deployment time and require a
specific configuration of the system surface during the deployment steps and imme-
diately upon the deployment.

The understanding of the aforesaid aspects has been gained from theoretical and
experimental investigations contributing each other. It should be remarked that
early in the deployment the solutions proposed by experienced designers and being
apparently obvious at first sight, further turned to be radically unacceptable or low
effective. The reason was that the object under study was, as a matter of fact, the stability of a flexible gyroscope possessing the specific inherent dynamics and, actually,
having no worldly analogues.

### 3.1.2. Folding Patterns for Film and Tether Structures

A film reflector folded into the bellows-type pattern (Fig. 3.1) is an example of a
structure which looses its stability while being deployed by centrifugal forces when
the film sheet is first folded along the axis and two ends of radial tethers are reeled
up on the central cylinder thereafter. The assumed deployment would have to occur
in a reverse order: after the central cylinder is set in rotation and holding links are
removed, the radial tethers would be spaced apart, each in its direction. Once the
tethers have been completely unreeled and positioned along one axis, the sheet
halves would be unfolded in different directions.

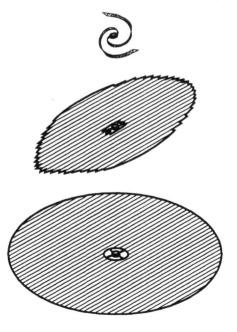

Fig. 3.1. Folding of the flexible
sheet into the bellows-type pattern

Investigators at the Central Research Institute for Machine Building and RSC Energia observed a quite different behaviour of film sheets of $D = 5$ m in the process of their deployment from the given folding pattern in large vacuum chambers. In the first half of the experiments the radial tethers stuck into one bundle in the process of unreeling and the deployment was not even initiated; in the second half of the experiments the tethers were unreeled separately, however, on completion of the unreeling, they were positioned not along one axis, but at a certain angle to each other, with both halves of the sheet being expanded in one direction. The process was uncontrolled and proceeded "in bursts". Investigation of the folding pattern as per ref. [2] (Fig. 3.2) showed a stable deployment procedure.

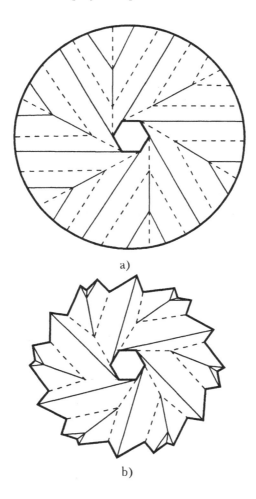

a)

b)

Fig. 3.2. a) Folding pattern; b) a folding pattern as per ref. [2] in the phase of deployment

Based on the experimental research analysis, a conclusion was drawn which high-lighted the necessity for employing the multi-beam folding patterns capable to main-tain stability in the field of centrifugal forces at the expense of their multi-link nature.

In designing a solar sailing vehicle, the problem was brought of how to simultane-ously deploy two counter-rotating systems and to control the deployment process regardless of centrifugal forces. By the controlled deployment is meant a process whose dynamic parameters could be controlled, modified, or specified under a cer-tain law. The process involves the deployment velocity control and a law of torque application. At the second step the bellow-type pattern deployment was proceeding "in bursts" and the process was uncontrolled. In the case of the folding pattern as per ref. [2] the deployment velocity relied on the driving shaft speed (twice as high). It was proposed to control the deployment from this pattern by introducing pressing rollers along the radial generatrices (Fig. 3.3). By controlling the running speed of the pressing rollers a completely controllable deployment could be achieved. How-ever, the folding pattern as per ref. [2] appeared to be technologically inefficient and extremely arduous *per se*. Besides, the attempt to use the pressing rollers created the problem of how to make the roller running over a thick, soft pattern. This was ac-companied by wrinkles and relative shifts of the package. To assure the structure from these effects it could have an underlay of crease-resistant film shaped as a nar-row beam along the pattern radius. However, this would add the complexity to the film production flow and the folding procedure which is fairly complicated per se. For this reason, and also because of the possible rotation of a blade relative to the ra-dial axis, further studies were focused on the film sheet structure consisting of sepa-rate sectors combined around the periphery (Fig. 1.2). Each of the sectors was folded into the bellows-type pattern in the same fashion as the pattern as per ref. [2], not tangentially, but radially directed, and reeled up on its spool. This folding pattern is derived from the folding pattern of separate strips, the concept described in the "he-liorotor" project.

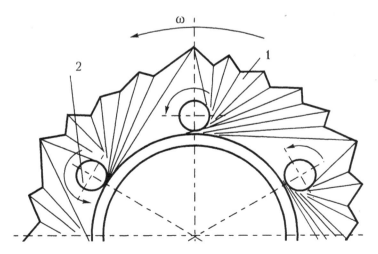

Fig. 3.3. Pressing rollers on the pattern as per ref. [2]. 1 — folding pattern; 2 — pressing roller

The process of deployment from spools combined with one driving mechanism had appeared to be completely controllable and was verified during the Znamya hardware ground tests and in-flight demonstration.

Further works have been conducted on the folding pattern types for solid reflectors offering a maximum performance of the reflecting surface and supposed for use in the space-based systems having as their purpose the illumination of the Earth's regions by the reflected sunlight.

The proposed Lamp concept was the first to incorporate the solid sheet folding pattern based on the Dolgoprudni Automatics Design Office (DADO) layout (Fig. 3.4). In the process of deployment by centrifugal forces the pattern is completely controllable through controlling the extension velocity of restraining tethers. Let us give a more detailed consideration to the folding pattern and show how its stability loss is attributed to possible modifications. This pattern falls into types of patterns being folded from the periphery. In the process of fabrication the sectors, being folded into the bellows-type pattern and joined together, are first pulled to the central cylinder (Fig. 3.5). No gas entrapping by the bellows-type pattern element is observed in closed volumes during outgassing in vacuum. The resulting radial beams can be further pulled to the central cylinder in different manners: the beams can be

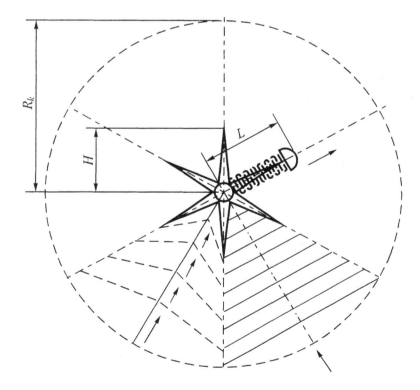

Fig. 3.4. Solid sheet folding into the DADO pattern. Step 1 of the reflector deployment (deployment of beams)

reeled up in spiral on the central cylinder (Fig. 3.6), folded into a "serpentine" pattern (Fig. 3.7), or reeled up on spools (Fig. 3.8). The pattern specific feature is that the two-step deployment is required: first, the beams are expanded and the sectors are deployed thereafter. If the spools containing tethers and combined by one driving mechanism were accommodated on the central cylinder with ends of the tethers attached to exterior spools, or the tether were fed through the "serpentine" pattern and pulled to the central cylinder, the controlled deployment of beams could be initiated first. Thereupon a similar assembly of spools could be used to deploy sectors. For this, the tethers have to be fed through or thrown over the pulled sector and fastened around its periphery. The possible option where the system looses stability is discussed thereafter. Suppose, we have done away with the sector tethers and are trying to deploy the system in one step by unreeling the tethers which are holding

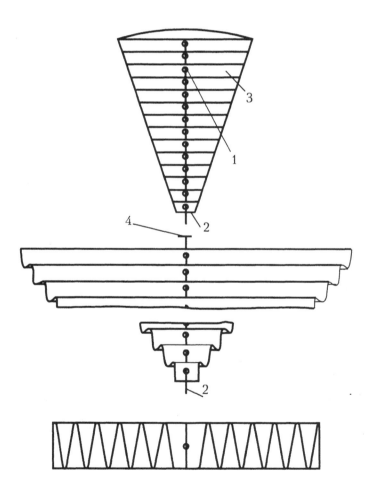

Fig. 3.5. Pulling up of the DADO sectors folded into the bellows-type patterns. 1 — hole; 2 — control tether; 3 — sheet sector; 4 — restrainer

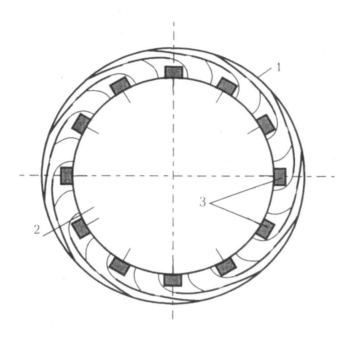

Fig. 3.6. Reeling up into a spiral of the DADO pattern beams. 1 — sheet beams; 2 — central cylinder; 3 — separators

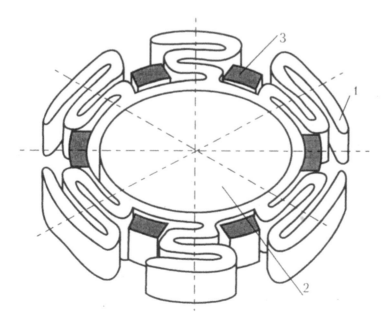

Fig. 3.7. "Serpentine" folding of the DADO pattern beams

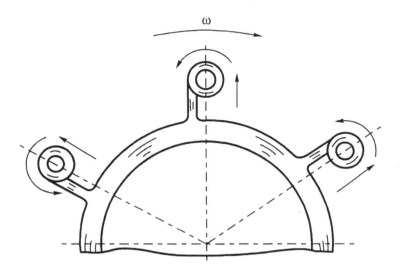

Fig. 3.8. Reeling up of the DADO pattern beams on spools

the beams reeled up on the spools (Fig. 3.8). Here, as the beams are being fed out from the spools, mass of film sectors distributed around the periphery of the structure being expanded will be fed out at the expense of centrifugal forces. Because the film is easily fed out from the exterior spools, the moment will rapidly occur when the central mass of the sectors will spontaneously commence to pull to the periphery while being fed out from the exterior spools irrespective of the tethers controlling the deployment from the spools and restricting a position of the exterior spools, not their possible rotation. Owing to irregularities of the structure, the sectors will be expanded at different times thus leading to the structure partial deployment "in bursts" when the major portion of mass of the film sectors will be at the maximum distance away from the central cylinder, i. e. will come to a position restricted by the length of the already expanded radial tethers.

When the structure incorporates no tethers attached to the sectors and the beams are folded into the "serpentine" pattern containing a tether fed through and attached to the beam periphery, the deployment of the sectors would be practically impossible. The tether fed through the "serpentine" pattern would inhibit the deployment of the sector mass.

Friction forces occurring in the "serpentine pattern" with the tether fed through would be beyond the control. Sticking would be expected in different beams throughout the deployment. If an exterior "bag" is put on the "serpentine" pattern, the situation would not be different from that when the stability is lost by the system with the beam reeled up on a spool, i. e. the film would be spontaneously fed out from the "bag" under the action of centrifugal forces applied to the main sector mass.

A rather stable deployment could be achieved throughout all steps by the sequential control of radial and sector tethers, the radial tethers being controlled first. Here, at the first step of deployment, in all three cases (the "serpentine" pattern containing a tether fed through, beam on a spool, beam in a "bag") a certain degree of stability would be observed. In the case with the tether fed through the beam eyelet no strong friction resistance would occur due to the lack of strong tangential forces. All steps of deployment should be undoubtedly subjected to the rigorous ground verification.

One more important requirement is a large number of sectors (actually equal to 8 or more). With the tether deploying mechanism failed or the film settled, the stability could be lost even in the theoretically stable system. For example, if we have only 3 sectors and, while being deployed, one spool containing sector tethers is locked by any reason and the film is settled and not separated from the central cylinder, the forces would be distributed asymmetrically from the geometry viewpoint and the pattern would become unstable and shifted to one side. With a large number of sectors the remainder would entail the defective one.

Consideration should be also given to the system using a solid film sheet folded from the centre (Fig. 3.9) and deployed "in bursts". The system has been considered at the early stages of development. Being a merely theoretical concept, no application has been found for the system because the peculiarity of the production flow required for large film sheets is that the sheets have to be folded as being fabricated. There is no way to fabricate a sheet of 100—200 m in diameter, spread it in a shop, and then fold into any pattern. The deployment "in bursts" also raised a problem because, in a relative motion, with film layers got into a friction, scores would inevitably appear on seams thus deteriorating the reflective coating.

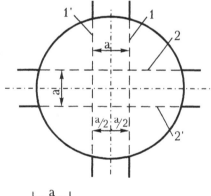

Fig. 3.9. Folding "from the centre" with unfolding "in bursts". 1, 2 — bend lines

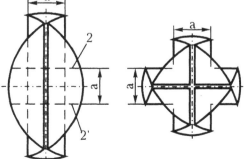

The aforesaid types of folding patterns had been discussed and investigated during the design studies and the selection was made against a much larger number of options.

To sum the investigated folding patterns they can be classified proceeding from the following features:

- split, solid,
- combined, non-combined around the periphery,
- folded from the centre or from the periphery,
- reeled up on separate spools or on the central cylinder using the coil or "serpentine" patterns,
- completely controllable, partially controllable, or uncontrollable.

The preference is given to this or that folding pattern depending on the application specific task. For example, in the "geliorotor" design [1] the solar sail is fabricated of separate strips which are not combined around the periphery. In this case the problem of the structure stability was not considered and, when the strips were fed out from spools asymmetrically, only the unbalance might be expected. On the contrary, in the Znamya-2 experiment the stable 8-beam pattern has been selected for the sectors combined around the periphery. When being deployed in space, three of eight sectors were settled because of a durable stowage and were deployed only at the end of the deployment process under the action of other sectors owing to the stability of the system, as a whole.

Experimental and theoretical investigations of the tether system deployment contributed much to the understanding of how the transformable systems are deployed (including film sheets). Possible structures of the circular tether system are shown in Fig. 3.10. The structures differ in a manner of fastening the circumferential contour to radial links. The possibility of loosing the shape stability during the structure deployment have been theoretically investigated. The conclusions of these investigations suggested that, to make the sliding structure stable, weights having a mass being in total less than the mass of tethers by a factor of 3.14 should be placed in the movable attachment points. 24 configurations of tether structures were experimentally investigated in the open air and in the most cases a loss of stability was observed during deployment. The results proved the conclusion on the stabilising effect of additional weights. It followed that only one structure (Fig. 3.10a) containing originally hard fastened tethers would be the most stable, simple, and efficient. The additional weights should have added to the system stability and, hence, the further work (on the tethered antenna development) was based exactly on this approach.

### 3.1.3. Torque Application Techniques

Theoretically, the torque could be applied both to the driving central cylinder and to the structure periphery. Suppose that a pair of jets is accommodated diametrically opposite on the film reflector being deployed, or on the circular tethered antenna. However, such a layout would require the jets to have their own attitude control system to maintain the jet motion within the plane of rotation (pitch angle stabilisa-

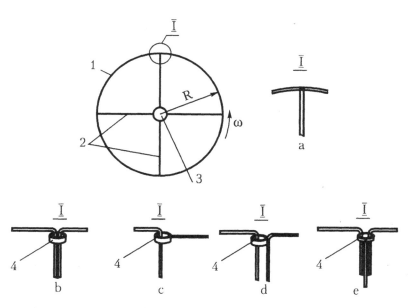

Fig. 3.10. Tether system structure. 1 — periferal tether; 2 — radial tether; 3 — central cylinder; 4 — sliding ring

tion). In the aforesaid "geliorotor" project [1] the solar sail made of separate strips has been deployed by the sunlight pressure acting on the expanded strips over the entire length. It was assumed to implement the blade "propeller effect" through its rotation around the radial axis. This option would be feasible if the blade could be sufficiently rigid or the deployment could be commenced in another way (in [1] plasma thrusters were employed). The torque application to the central cylinder would be the most versatile. In principle, the torque could be governed by any law using a feedback (a tracking drive). Let us consider the possibility of applying the torque generated by different mechanisms such as a flywheel run to a high speed a priori, the powder or gas jets (with the thrust being changed under different laws depending on time), electric drives with rigid or drooping characteristics of electric motors. The use of the flywheel accumulated kinetic energy has been proved unreasonable because, while the system is being deployed by centrifugal forces, not the kinetic energy, but the angular momentum would be stored. There, the surplus kinetic energy should be removed from the system in the process of deployment. Otherwise, the energy excess could lead to undesirable oscillations and the system failure. Actually, based on the law of conservation of kinetic energy the following expression can be written:

$$K = J_0 \omega_0 = J_k \omega_k \qquad (3.1)$$

where $J_0$ and $\omega_0$ — moment of inertia and angular velocity of the flywheel,
$J_k$ and $\omega_k$ — moment of inertia and angular velocity of the reflector.

In the case with a steel flywheel of radius $r = 0.1$m and angular velocity $\omega_0 = 10^2$ rad/s, the moment of inertia is:

$$J_0 = \frac{1}{2} m r^2 = \frac{10^2 (10^{-1})^2}{2} = 0.5 \text{ kg} \cdot \text{m}^2 \tag{3.2}$$

The angular momentum is:

$$K = J_0 \omega_0 = 0.5 \cdot 10^2 \frac{\text{kg} \cdot \text{m}^2}{\text{s}}$$

Here, the kinetic energy of the flywheel will be:

$$E_0 = \frac{J_0 \omega_0^2}{2} = \frac{0.5 \cdot 10^4}{2} = 0.25 \cdot 10^4 \text{ J} \tag{3.3}$$

The moment of inertia of the film reflector of radius $R_k = 100$ m and mass $m = 250$ kg is:

$$J_k = \frac{1}{2} M R_k^2 = \frac{2.5 \cdot 10^2 \cdot 10^4}{2} = 1.25 \cdot 10^6 \text{ kg} \cdot \text{m}^2.$$

The reflector final angular velocity derived from equation (3.1) will be equal to:

$$\omega_k = \frac{J_0 \omega_0}{J_k} = \frac{0.5 \cdot 10^2}{1.2 \cdot 10^6} = 0.5 \cdot 10^{-4} \text{ rad/s} \tag{3.4}$$

With such an angular velocity the reflector kinetic energy will run only at:

$$E_k = \frac{J_k \omega^2}{2} = \frac{1,2 \cdot 10^6 (0,5 \cdot 10^{-4})^2}{2} = 1,15 \cdot 10^{-2} \text{ J}$$

Comparing kinetic energy $E_0$ accumulated in the flywheel with kinetic energy $E_k$, we see that $E_k$ makes a negligibly small part of $E_0$ and, actually, the total amount of $E_0$ should be removed from the system in any way. The feasibility of connecting the rotating flywheel to the film system being deployed is also problematic.

It would be quite a different matter if the system, throughout its deployment, could be used in the capacity of the flywheel with the accumulated angular momentum being already imparted to the system in any way. If the stowed system is first spun and then deployed, the situation would be such as in the foregoing example with the flywheel, that is the surplus energy must be removed and, with the maximum possible initial velocity, the final velocity would be low in any case and might not meet the application specific task requirements. However, if the system had been otherwise partially deployed, the deployment process could be efficiently completed thereupon at the expense of the originally accumulated angular momentum. This situation would be encountered when the tether systemwas being deployed stepwise.

When using the gas or powder drive, one essential feature would be observed comparing to the electric drives. This is the angular momentum accumulation occurring while the system is being deployed with no requirements for its compensation by introducing a counter-rotating system. The theoretical solution of the deployment dynamics task with the constant or time variable angular momentum suggests the oscillatory nature of the deployment. The ground verification would brings difficulties associated with the requirements for maintaining a sufficient level of vacuum in a chamber with the gas jet in service. There, it seems likely that the ground verification would require masses to simulate the film sheet. The jets would act on the sheet changing its shape.

When the electric drive is employed, the angular momentum of the system being deployed has to be compensated. That is, if the electric drive rotor is combined with the system being deployed, its stator would rotate in the opposite direction and should be combined with a system capable to compensate the angular momentum. When the reflector has been deployed on the Progress vehicle, the moment of inertia of the vehicle was beyond that of the deployed film sheet by a factor of 40. The vehicle body connected to the electric drive stator was slowly rotating in the direction opposite to the reflector rotation. Here, the attitude control system incorporated the control system chemical fuel thrusters arresting the rotation.

In the case with a solar sailing vehicle (Fig. 1.2) the angular momentum of the reflector is compensated by a counter-rotating tether flywheel or the second reflector and the gyroscopic forces resulting from the deflection of axis of two counter-rotating structures are employed to perform the system attitude control.

With the rigid characteristic of the electric drive, that is when the angular velocity is actually constant depending on the generated momentum, the following drawbacks would be found (Fig. 3.11, a):

1) To make the sheet strong at the end of the deployment process the entire process should proceed exactly at this minimal angular velocity thus increasing the deployment time; and, owing to small centrifugal forces, it would be difficult to initiate the deployment.

2) The calculations showed the oscillatory nature of the deployment dynamics, that is during the deployment the periodic, partial reeling off — reeling up of the reflector was observed that would be impermissible because the sheet would be fed out from the production pattern and, with no special guides, the film sheet could be reeled up randomly; at the end of deployment a tendency for the reverse reeling up of the sheet would be expected (this could be prevented by installing a freewheel clutch).

By using the drooping characteristic of the electric drive (Fig. 3.11, b), that is, when the angular velocity is reduced as the moment is increased, the aforesaid drawbacks occurring at the rigid characteristic would be eliminated and the following advantages would be offered:

1) At the initial step of deployment the angular velocity could be sufficient to initiate the deployment process.

2) A rather high velocity of deployment and a short time of deployment could be provided.

3) The calculations has shown that the deployment would be smooth at all steps with no oscillations of the reeling off-reeling up type at the intermediate and final steps of deployment.

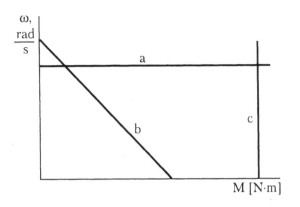

Fig. 3.11. Principal characteristics of electric drives

4) At the end of deployment a low velocity of rotation could be achieved in response to the system material strength.

In this case the self-controlled, stable deployment of the system is observed. The stability throughout the deployment is governed exactly by the specific, drooping characteristic of the drive. When the possible, incidental departure from the system dynamics occurs, the system would be recovered into a stable state at the expense of the above characteristic of the drive. For example, if, for any reason, the system was slowed down, this would result in reducing the drive speed and increasing the torque which, in turn, would run the system again; on the contrary, if the system was accelerated, this would result in decreasing the torque and braking the system.

Similar processes happen in a number of self-controlled systems. For example, a welding arc is steadily burning depending on the dc generator. Any incidental deviation of the arc parameters would cause the generator electric parameters to change thus recovering the arc state. The above advantages of the electric drive possessing the drooping (falling) characteristic provided a basis for the author's certificate. The d.c. motors widely used in the engineering technology, including space technology, have this drooping characteristic. Among the advantages offered by the electric drives based on the d.c. motor are their reliability, easiness of ground verification, good compatibility with other systems of the vehicle such as control, nutrition, housekeeping systems (explosion-proof, no gas release), etc.

The electric drive possessing the drooping characteristic has been successfully employed in the space-based Znamya-2 experiment and incorporated in all the follow-on projects.

For complicated systems requiring to be deployed stepwise, theoretically, a tracking drive could be designed which would obey any law of parameters variation and have the feedback provided via sensors and the onboard computer. However, many of the modes could be obtained by using the freewheel and torque clutches (in the Znamya-2 experiment both clutches have been used to enhance reliability).

In calculating the reflector deployment dynamics the parameters to be defined are characteristics of the reflector electric drives responsible for the rotation and deployment processes and making stable the geometry of the structure being deployed throughout all steps and further operation. The numeric calculations provides a basis to establish requirements for electric drives. The law of applied torque and the reflector deployment velocity should be selected so that the deployment is assured during a required time period with no undesirable torsional oscillations and entanglement of the flexible system. The electric drives should run the central cylinder carrying the reeled up reflector to the initial velocity to initiate the deployment and maintain its rotation throughout the operation at the angular velocity dependent on the film material strength.

### 3.1.4. Ways of Providing Controllable Deployment

In parallel with the torque application to the system being deployed, the system should be capable to change its dimensions. Otherwise, not the deployment, but the acceleration of the system would occur. The process of changing the system dimensions could be uncontrolled, partially controlled, and completely controlled. The process of deploying the tethered weights is serving as an example of the uncontrolled deployment when the tethers are reeled up on freely rotating spools accommodated on the rotating central cylinder. Once the central cylinder is set in rotation and the retaining links are removed, the weights would come apart tangentially to the central cylinder. With the limited length of the tethers an impact would occur at the end of the process and, if the system is not destroyed, the tethers would be reeled back on the central cylinder (not on the spools). The process of unreeling the tethered weights from the central cylinder presents an example of the partially controlled deployment. The deployment velocity depends on the central cylinder speed. At the end of the process, if the freewheel clutch is not provided, the tethered weight would be reeled back on the central cylinder. The process with the forced unreeling of the tethered weights from spools accommodated on the central cylinder and combined via their drive is an example of the completely controlled deployment. To prevent the system from getting into a tangle the deployment drive speed should not be arbitrary ("getting into a tangle" means that elements of the system being deployed are deflecting from the radial direction at an angle greater than 90°). Suppose that the tethered weights are fed out from spools extremely fast with the central cylinder rotating at a relatively low speed. And, hence, it appears that the centrifugal forces could be insufficient to expand the tethered weights and the weights would loose their link with the driving centre, that is the system would get into a tangle. To make the process stable the centrifugal force, $F_{cf} = m \omega^2 R$, expanding the tether in the radial direction should be well beyond the Coriolis force acting laterally, $F = 2 m V \omega$ , (where $V$ — deployment velocity) and the inertial force, $F = m \dot{\omega} R$ , irrespective of the law of the applied torque, viz :

$$m \omega^2 R >> 2 m V \omega \quad \text{or} \quad \frac{\omega R}{2 V} >> 1 \tag{3.5}$$

$$m \omega^2 R >> m \dot{\omega} R \quad \text{or} \quad \frac{\omega^2}{\dot{\omega}} >> 1 \tag{3.6}$$

Actually, the above conditions are easy to implement and the deployment velocity is selected while obtaining the mathematical solution on the deployment dynamics with the required margin for the angle of deflection from the radial direction. The deployment velocity could be constant, smoothly or stepwise changing. By reducing the unfolding velocity during deployment the process would be stabilised, however, this would inhibit from reducing the angular velocity of rotation to a value sufficient from the strength or other considerations. For example, in the space-based Znamya-2 experiment, the change of the spool radius in the process of the film unfolding resulted in the 2-fold reduction of the deployment velocity that prevented from the reduction of the central cylinder rotation velocity and transition to the second (low speed) characteristic of the drive. An additional resistor had to be introduced in the primary drive power circuit to increase the slope of its characteristic.

In solving tasks of deployment dynamics for complicated tether systems being at the limit of stability, the options are found when the system is deployed to a certain state, whereupon the deployment is terminated and the system is run to accumulate the margin for stability, and, from there, the deployment is proceeding at a lower velocity. The conditions described in (3.5, 3.6) are insufficient to achieve the required stability.

Summing up this Chapter, we might reason that the process dynamics of deploying the stowed structures by centrifugal forces depends on such factors as a folding pattern, an applied torque, and a deployment velocity. To obtain a stable package geometry, in most cases the modes could be selected through mathematical simulation to provide the system deployment.

We have to emphasise that one of the current tasks is to describe theoretically the process of deploying large space-based structures by centrifugal forces and to elucidate and employ specific features attributed to centrifugal forces. The field of centrifugal forces is of a central nature and prevents the rotating system elements from possible offsets both from the rotation plane and from the radial direction within the rotation plane.

In selecting the deployment modes, the process dynamics should be such as to make centrifugal forces predominant over the other dynamic constituents, thus ensuring the stabilising effect of centrifugal forces. As a consequence, this enables to develop the relatively simple software and experimental verification procedures based on the requirements (3.5, 3.6) to be implemented. The validity of this approach has been proved by comparing the calculation techniques given below with the results of the in-orbit and ground experiments (see Chapter 7).

## 3.2. Mathematical Model for Split Reflector Controllable Deployment

Consider now the mathematical model of the split reflector deployment, the reflector consisting of separate sectors combined around the periphery. The sectors are arranged following the fan-like pattern (Fig. 3.12), while the folds are arranged along the sector radial generatrices and reeled up on separate spools accommodated on a central cylinder, or on one common spool. The central cylinder together with the accommodated spools is set in rotation and the controllable deployment of the reflector sectors is initiated from the spools hereafter following the sought-for law. The analysis takes account of a system consisting of a central cylinder, flexible elements of the reflector, and drives possessing characteristics responsible for the law of change of the torque applied to the central cylinder and the law of film deployment speed change when the film sheet is being fed out from a spool. Since the motion of all sectors in the model is taken equal, the Equation of motion is written only for one sector with the total value for all sectors being summarised when the force applied to the central cylinder from the sectors is recorded.

At first, let us consider the mathematical model of the system consisting of a central cylinder and an unstretchable tether with a weight attached to its end in so far as the model is physically descriptive on one hand and possesses many properties of the distributed mass systems on the other.

Basic assumptions taken for the development of the mathematical model of the given system are:
   • the system is being deployed symmetrically relative to the central axis,
   • radial tethers are straight,
   • relative transitions in the system are going on within a plane normal to the system axis of rotation,
   • the system energy dissipation caused by deformation, friction, and environmental effects is neglected.

Introduce the right orthogonal baselines (Fig. 3.12): inertial baseline $\bar{e}$ and free baselines $\bar{e}^{(1)}$ and $\bar{e}^{(2)}$. Put the start of baseline $\bar{e}$ in the centre of system 0, direct $\bar{e}_3$ along the axis of rotation, then $\bar{e}_1$ and $\bar{e}_2$ will lie within the system deployment plane. Hardly refer baseline $\bar{e}^{(1)}$ to the central cylinder and also place its start in point 0, match axis $\bar{e}_3^{(1)}$ with $\bar{e}_3$. Start baseline $\bar{e}^{(2)}$ put from point A, where the radial tether comes out, direct vector $\bar{e}_3^{(2)}$ similar to $\bar{e}_3$, direct vector $\bar{e}_1^{(2)}$ along the radial tether.

Angle of deflection from the radial direction, $-\varphi_2$, will be further designated $\varphi$ to simplify the equation.

Now, the radius-vector of the weight is:

$$\bar{R} = \bar{r}_0 + \bar{L} \tag{3.7}$$

where $\bar{r}_0$ — radius of the central cylinder,
   $\bar{L}$ — radius-vector of the weight in baseline $\bar{e}^{(2)}$ (the length of the deployed portion of tether).
   Then:

$$\dot{\bar{R}} = \bar{\omega} \times \bar{r}_0 + \overset{v}{\bar{L}} + (\bar{\omega} + \dot{\varphi}\,\bar{e}_3^{(2)}) \times \bar{L} \tag{3.8}$$

where $\bar{\omega}$ — angular velocity of the central cylinder relative to the inertial system;
   "v" designates a local derivative in baseline $e^{(2)}$

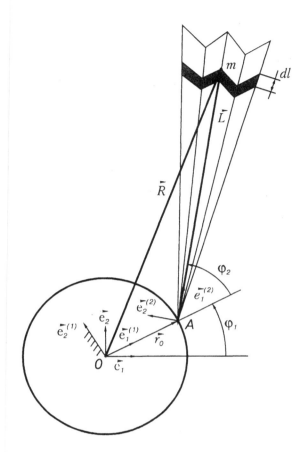

Fig. 3.12. Coordinate systems of the split reflector controlled deployment

$$\ddot{\bar{R}} = \dot{\bar{\omega}} \times \bar{r}_0 + \bar{\omega} \times (\bar{\omega} \times \bar{r}_0) + \ddot{\bar{L}} + 2(\bar{\omega} + \dot{\varphi}\, e_3^{(2)}) \times \dot{\bar{L}} +$$

$$+ (\dot{\bar{\omega}} + \ddot{\varphi}) \times \bar{L} + (\bar{\omega} + \dot{\varphi}\, e_3^{(2)}) \times ((\bar{\omega} + \dot{\varphi}\, e_3^{(2)}) \times \bar{L})). \tag{3.9}$$

Write the basic equation of dynamics for the weight as:

$$m\,\ddot{\bar{R}} = F \tag{3.10}$$

In projections on the baseline axis $\bar{e}^{(2)}$, using (3.9), we obtain:

$$m\left[r_0(\omega^2\cos\varphi - \dot{\omega}\sin\varphi) - \ddot{L} + L(\omega + \dot{\varphi})^2\right] = N, \tag{3.11}$$

$$r_0(\dot{\omega}\cos\varphi + \omega^2\sin\varphi) + 2(\omega + \dot{\varphi})\dot{L} + (\dot{\omega} + \ddot{\varphi})L = 0, \tag{3.12}$$

where $N$ — stress over a root cross-section,
  $\dot{L}, L$ — sought-for laws of the tether deployment control.
  Express the theorem of a angular momentum variation for the central cylinder as:

$$\dot{\overline{K}}_1 = \overline{M} + n\,(\overline{r}_0 \times \overline{N}),\qquad (3.13)$$

where $M$ — sought-for control law of change of external torque applied from a drive,
  $n$ — a number of tethered weights.
  By projecting to the axis of rotation we obtain:

$$J\,\dot{\omega} = M + n\,N\,r_0 \sin\varphi \qquad (3.14)$$

where $J$ — moment of inertia of the central cylinder.
  Combined equations (3.11, 3.12, 3.14) describe the deployment of the mechanical system consisting of a central cylinder, tethered weights, and control systems possessing certain characteristics.
  To obtain the equations of motion for a reflector sector, let us write the sector element mass

$$dm = \frac{2\pi\beta}{n}\,(l + L_k - L)\,dl,$$

where $l$ — current length of the deployed sector,
  $L_k$ — final length of the sector,
  $\beta$ — density of the film surface.
  Replacing $m$ with $dm$ in (3.10) and integrating along the deployed portion of the sector in projections on the axis of baseline $\overline{e}^{(2)}$ we have:

$$\frac{2\pi\beta}{n}\,(L_k L - \frac{L^2}{2})\,(r_0(\omega^2\cos\varphi - \dot{\omega}\sin\varphi) - \ddot{L}) +$$

$$+\frac{2\pi\beta}{n}\,(\frac{L_k L^2}{2} - \frac{L^3}{6})\,(\omega + \dot{\varphi})^2 = N; \qquad (3.15)$$

$$(L_k - \frac{L}{2})\,[\,r_0(\dot{\omega}\cos\varphi + \omega^2\sin\varphi) + 2(\omega + \dot{\varphi})\dot{L}\,] +$$

$$+(\frac{L_k L}{2} - \frac{L^2}{6})\,(\dot{\omega} + \ddot{\varphi}) = 0 \qquad (3.16)$$

where $N$ — tension in the sector root cross-section.
  The equation for the central cylinder is similar to (3.14).
  Combined equations (3.14, 3.15, 3.16) describe the deployment of mechanical system consisting of a central cylinder, a reflector, and control systems having the required characteristics. With respect to this system, the possibility of a proper control has been studied. Taken as control parameters were: $M$ — external moment of the drive, $L$ — deployment velocity of the flexible link.

This system of equations reduced to the Cauchy's form was integrated using the method of the forth order under initial conditions with $\omega(0) = \omega_0$, $\varphi = \dot{\varphi} = 0$.

Studies have been made on the deployment stability of the mechanical system consisting of a film reflector, a central cylinder, and a rotation drive with characteristics depending on stability parameter $\gamma = L\omega/(2V)$ ($V$ — deployment velocity) and on how the drive characteristics are varying under law $M = M_0(1 - \omega/\omega_0)$ at values $K = M_0/\omega_0 = 0, 0.1, 1, 3, 10, \infty$. Stability parameter $\gamma$ describes the relationship between centrifugal forces keeping the system stable and the Coriolis forces causing the system to tip out from the radial direction. The dependence of the angle of deviation on parameter $\gamma$ is shown in Fig. 3.13 for the system with $\beta = 10^{-3}$ kg/m$^2$, $R_k/r_0 = 66.7$, $\omega_0 = 10$ rad/s, for two values of deployment velocity, where $1 - K = 0.1$, $V = 0.2$ m/s; $2 - K = 1$, $V = 0.2$ m/s; $3 - K = 3$, $V = 0.2$ m/s; $4 - K = 10$, $V = 0.2$ m/s; $5 - K = 0.1$, $V = 0.05$ m/s; $6 - K = 3$, $V = 0.05$ m/s; $7 - K = 3$, $V = 0.05$ m/s, $R_k/r_0 = 200$; $8 - K = 3$, $V = 0.05$ m/s, $R_k/r_0 = 666.7$. Modes with $M = K\omega$ are unstable beginning from the initial stages.

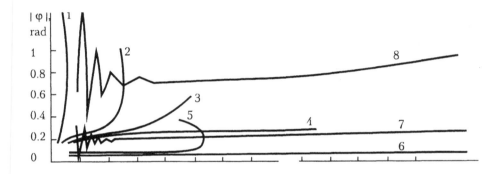

Fig. 3.13. Angle $\varphi$ versus dimensionless parameter $\gamma$ during deployment

The diagrams show that the least value of the angle of deviation corresponds to values with $K = 1-10$. It follows from diagrams 6, 7, 8 that changing $R_k/r_0$ from 67 to 670 caused $\varphi_2$ to increase from 0.07 to 0.7 rad.

Diagrams 1 and 2 demonstrate that at $K = 0.1-1$ (modes with external moment values being constant in time or modes similar to them) the oscillatory deployment of the split reflector would be expected.

## 3.3. Mathematical Model of Solid Reflector Controllable Deployment

For purpose of solving a number of tasks it would be reasonable to use a solid film canvas (e.g. for detection of space debris, illumination of the Earth regions with reflected sunlight in the night time, recovery of ozone layer by accommodating a sunlight reflector on the sun-synchronous orbit, etc.).

Consider the mathematical model for the controlled deployment of solid film reflector folded into the DADO-type pattern similar to that discussed in par. 3.2 (Fig. 3.4).

Basic assumptions taken in creating the mathematical model of the given system:
- the system is being deployed symmetrically relative to the central axis,
- the radial components are straight,
- relative motions within the system are going on in the plane normal to the system axis of rotation,
- the system energy dissipation caused by strain, friction, and environmental effects is neglected.

The split reflector is deployed in two steps: initially, the controlled deployment of the reflector radial part $(n)$ $H = 2 \pi R k / 2 n$ long (see Fig. 3.4) takes place and the remainder of the reflector is deployed at the second step. (See Fig. 3.14).

Consequently, mathematical simulation of the reflector deployment process is also performed in two steps.

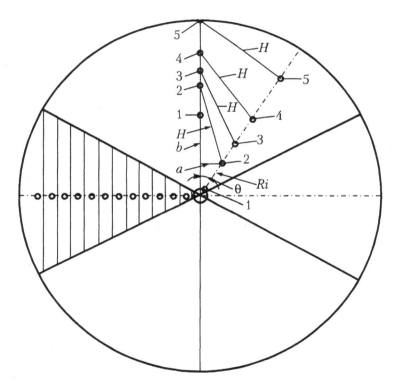

Fig. 3.14. Solid reflector deployment step 2 (sheet deployment)

Step I: Simulated deployment of beams $n$. During simulation the beams are replaced with tethers of linear density $\mu$ containing weights attached to their ends,

$$\mu = \frac{m_{ref}\, k}{H\, n}\,,\ \ H = \frac{\pi\, R_k}{n} \tag{3.17}$$

where  $m_{ref}$  —  reflector mass,
$\quad\quad n$  —  number of sectors,
$\quad\quad k = 1.2$  —  factor of the reflector mass increase due to the operations relative
$\quad\quad\quad\quad\quad\quad$  to the film initial mass.

Mass of weights attached to the ends:  $m_w = H\mu - L\mu$ .
Hence, the equation for the central cylinder will be written as:

$$J\,\dot\omega = M + n\, N\, r_0 \sin\varphi\,, \tag{3.18}$$

where  $M$  —  exterior moment of the drive,
$\quad\quad J$  —  moment of inertia of the central cylinder,
$\quad\quad N$  —  film tension in the root cross-section,
$\quad\quad r_0$  —  radius of the central cylinder,
$\quad\quad \omega$  —  angular velocity of rotation of the central cylinder,
$\quad\quad \varphi$  —  angle of deflection from the radial direction.

Equations of dynamics for the sector will be written as:

$$N = n\,\mu\, \dot H\,[\,r_0\,(\omega^2 \cos\varphi - \dot\omega \sin\varphi) - \ddot L\,] + (\mu H - 0.5\, L\,\mu)(\dot\omega + \ddot\varphi) \tag{3.19}$$

$$H\,[\,r_0\,(\dot\omega \cos\varphi + \omega^2 \sin\varphi) + 2\,(\omega + \dot\varphi)\,\dot L\,] + (H - 0.5\, L)(\omega + \dot\varphi)\,\dot L = 0 \tag{3.20}$$

where  $L$  —  current length of the sector,
$\quad\quad \dot L$  —  sought-for control parameter governing the deployment velocity.

And, hence, we obtain the system of equations (3.18, 3.19, 3.20) describing the deployment dynamics of the split reflector at the first step.

Step II: At the second step, having the beams expanded, the remainder of the reflector is deployed (Fig. 3.14). To consider the mass of the expanded beams, an actual radius of the reflector is replaced at a sufficient accuracy:

$$\tilde L = L + 0.5\, b\,,$$

where

$$b = (H^2 - a^2)^{1/2}\,,\ \ a = L \sin\beta\,,\ \ \beta = \frac{360}{2\,n}\,.$$

The equation for the central cylinder is similar to (3.18). Equations (3.19) and (3.20) for the reflector will be written:

$$N = \pi\,\beta\, R_k^{\,2}\Big[\,r_0\,(\omega^2 \cdot \cos\varphi - \dot\omega \sin\varphi) - \ddot L\,\Big] +$$

$$+ \pi \beta \left( R_k^2 - \frac{L^2}{3} \right) \left[ L \left( \omega + \dot{\varphi} \right)^2 \right], \tag{3.21}$$

$$\pi \beta R_k^2 \left[ r_0 \left( \dot{\omega} \cos \varphi + \omega^2 \sin \varphi \right) + 2 \left( \omega + \dot{\varphi} \right) \dot{L} \right] +$$

$$+ \pi \beta \left( R_k^2 - \frac{L^2}{3} \right) \left[ L \left( \dot{\omega} + \ddot{\varphi} \right) \right] = 0 \tag{3.22}$$

Hence, we obtain combined equations (3.18, 3.21, 3.22) describing the deployment dynamics of the solid reflector at the second step.

The given equations are numerically solved by the Rounge–Kutta method of the forth order. Under the initial conditions $\omega(0) = \omega_0$, $\varphi(0) = \dot{\varphi}(0) = 0$.

The numerical calculation results for the controllable deployment of the solid reflector are illustrated by the example of the system containing 8 sectors with the central cylinder mass of 300 kg. As the result of the numerical calculation analysis, the 20 W electric motor of various reduction ratios, $M_1 = 15 \left( 1 - \omega/10 \right)$, $M_2 = 1500 \left( 1 - \omega/0.1 \right)$ and a deployment velocity varying from 0.1 to 0.075 m/s has been selected.

Figures 3.15, 3.16 show dynamic parameters of the reflector controlled deployment at the first and the second steps.

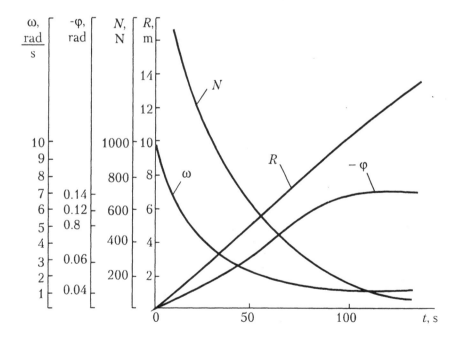

Fig. 3.15. Dynamic parameters of the solid reflector controlled deployment at step 1. $\varphi$ — angle of deflection from the radial direction; $\omega$— angular velocity of the central cylinder rotation; $N$ — total tension in the root cross-section; $R$ — structure radius

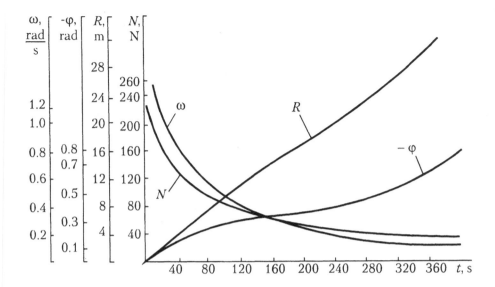

Fig. 3.16. Dynamic parameters of the solid reflector controlled deployment at step 2. φ — angle of deflection from the radial direction; ω — angular velocity of the central cylinder rotation; N — total tension in the root cross-section; R — structure radius

Figure 3.15 shows dynamic parameters for the first step of deployment: the deployment velocity varies from 0.1 to 0.75 m/s. The reflector has been smoothly deployed in 147 seconds.

Dynamic parameters for the second step of deployment are shown in Fig. 3.16. The deployment velocity varies from 0.1 to 0.075 m/s. The reflector has been smoothly deployed in 439 seconds. The final data on the deployment of the reflector beams served as the initial data for the second step.

The second procedure of the solid reflector deployment would be possible with the beams coiled on the central cylinder. This option renders unnecessary the controlled deployment at the first step that actually makes the structure twice as simple.

## 3.4. Mathematical Model of Solid Reflector Uncontrolled Deployment

To simulate the given process of the solid reflector deployment, let us give the first consideration to the "tethered weight" system where the tether is fed out from the central cylinder by inertial forces generated by the system rotation. This system differs from the aforesaid deployment process where the tether is fed out from spools under control in lacking the knowledge of the law of deployment a priori because it depends on the deployment process and is governed by the law of motion, as a whole.

Define the weight position as (Fig. 3.17):

$$\overline{R} = \overline{r}_0 + \overline{L},$$

$$\dot{\overline{R}} = (\overline{\omega} + \dot{\varphi}_2 \overline{e}_3^{(2)}) \times (\overline{r}_0 + \overline{L}) + \overset{v}{\overline{L}}, \tag{3.23}$$

where $\overline{\omega}$ — angular velocity of the central cylinder relative to the inertial system;

$\varphi_2$ — angle of deflection from the radial direction (further designated as $\varphi$);

$r_0$ — radius of the central cylinder;

$L$ — length of the deployed portion;

v — means the local derivative in basis $\overline{e}^{(2)}$.

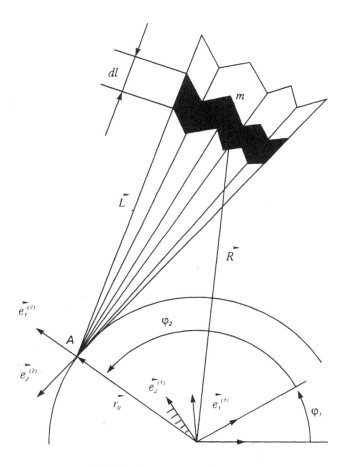

Fig. 3.17. Solid reflector controlled deployment coordinate systems (unfolding from the central cylinder)

The acceleration vector is obtained by differentiating

$$\ddot{\bar{R}} = (\dot{\bar{\omega}} + \ddot{\varphi}\,\bar{e}_3{}^{(2)}) \times (\bar{r}_0 + \bar{L}) + (\bar{\omega} + \dot{\varphi}\,\bar{e}_3{}^{(2)}) \times$$

$$\times \left[ (\bar{\omega} + \dot{\varphi}\,\bar{e}_3{}^{(2)}) \times (\bar{r}_0 + \bar{L}) \right] + \overset{vv}{\bar{L}} + 2(\bar{\omega} + \dot{\varphi}\,\bar{e}_3{}^{(2)}) \times \overset{v}{\bar{L}}. \qquad (3.24)$$

To derive the equation of motion, the principal equation of dynamics is projected for the weight on the axis of basis $\bar{e}_3{}^{(2)}$. In this way we obtain:

$$(\dot{\omega} + \ddot{\varphi})L + 2(\omega + \dot{\varphi})\dot{L} - r_0(\omega + \dot{\varphi})^2 = 0, \qquad (3.25)$$

$$N = m\,[\,r_0(\dot{\omega} + \ddot{\varphi}) + L(\omega + \dot{\varphi})^2 - \ddot{L}\,], \qquad (3.26)$$

where $N$ — tether tension;
    $m$ — weight mass.

Changes of an angular momentum for the central cylinder in projections to the system axis of rotation:

$$J\dot{\omega} = M - nNr_0. \qquad (3.27)$$

where $J$— moment of inertia of the central cylinder;
    $n$ — a number of weights;
    $M$ — external moment;
    $N$ — tension in the root cross-section;
    $r_0$ — radius of the central cylinder.

A trajectory of the weight motion relative to the central cylinder is an evolvent of the circumference of radius $r_0$ described by the equation:

$$L = r_0\,\varphi \qquad (3.28)$$

Hence:

$$\dot{L} = r_0\,\dot{\varphi} \qquad (3.29)$$

$$\ddot{L} = r_0\,\ddot{\varphi} \qquad (3.30)$$

By substituting (3.28—3.30) in (3.25, 3.26) we obtain:

$$(\dot{\omega} + \ddot{\varphi})\varphi + \dot{\varphi}^2 - \omega^2 = 0 \qquad (3.31)$$

$$N = m\,r_0\,[\,\dot{\omega} + \varphi(\omega + \dot{\varphi})^2\,] \qquad (3.32)$$

(3.27, 3.31, 3.32) are equations of motion for the system under consideration where the tether is rectilinear, weightless, and unstretchable.

To derive the equations of motion for the solid sheet let us turn to variable $L$ in (3.31, 3.32) and assume $l$ to be a current length of the deployed portion of the sector of $L$ long. The sector element mass is:

$$dm = \frac{2\pi}{n} \beta \, (l + L_k - L) \, dl, \tag{3.33}$$

where $\beta = \rho\,\delta$ — film surface density,
  $L_k$ — full sector long.

By substituting $dm$ for $m$ in (3.32) and integrating over the entire length of the deployed portion we have:

$$N = \frac{2\pi}{n} \beta \int_0^l [\, r_0\dot\omega + l(\omega + \dot\varphi)^2\,](l + L_k - L)\,dl =$$
$$= \frac{2\pi}{n}\beta L\left[\,(L_k - L/2)r_0\dot\omega + \frac{L}{6}(3L_k - L)(\omega + \dot\varphi)^2\,\right] \tag{3.34}$$

Similarly, from (3.31) we obtain:

$$\int_0^l\left[\,(\dot\omega + \ddot\varphi)\frac{l}{r_0} + \dot\varphi^2 - \omega^2\,\right](l + L_k - L)\,dl =$$
$$= \frac{L}{6r_0}(\dot\omega + \ddot\varphi)(3L_k - L) + (\dot\varphi^2 - \omega^2)(L_k - \frac{L}{2}) = 0 \tag{3.35}$$

Equations (3.27, 3.34, 3.35) are equations of motion for the sheet being fed out from one of the fan-like patterns, e.g. the pattern as per ref. [2] (Fig. 3.2).

When the DADO folding pattern is employed (Fig. 3.3), the reflector beams at the first step are replaced with the loop-shaped tethers which linear density is

$$\mu = \frac{m_{ref}\,k}{H\,n}\,, H = \frac{\pi R_k}{n}\,.$$

To derive equations for the beams, turn to variable $l$ in (3.31, 3.32). The tether element mass

$$dm = \frac{2\pi\mu\,dl}{n}$$

By substituting $dm$ for $m$ in (3.32) and integrating along the entire length of the released tether portion we have:

$$N = \frac{2\pi\mu}{n}\int_0^l\left[\,r_0\dot\omega + l(\omega + \dot\varphi)^2\,\right]dl =$$
$$= \frac{2\pi\mu}{n}L\left[\,r_0\dot\omega + \frac{L}{2}(\omega + \dot\varphi)^2\,\right] \tag{3.36}$$

Similarly, from (3.31) we have:

$$\int_0^L \left[ (\dot\omega + \ddot\varphi)\frac{l}{r_0} + \dot\varphi^2 - \omega^2 \right] dl =$$

$$= L\left[ (\omega + \ddot\varphi) L / (2 r_0) + \dot\varphi^2 - \omega^2 \right] = 0 \qquad (3.37)$$

(3.27, 3.36, 3.37) are equations of motion at the second step of deployment when the reflector beams are fed out from the central cylinder. The second step is described by Equations (3.18, 3.21, 3.22).

Combined equations (3.18, 3.21, 3.22) and (3.27, 3.36, 3.37) are solved by the Rounge—Kutta numerical integration method of the forth order. Initial conditions for the first step: $\varphi(0) = \dot\varphi(0) = 0$, $\omega(0) = \omega_0$; at the second step the initial conditions for variables $\varphi$ and $\omega$ are their values calculated at the first step.

Results of the numerical calculations of the solid reflector non-controlled deployment are analysed by using an example of the system containing a central cylinder of 300 kg mass and an electric motor possessing characteristics similar to those for the two-step controlled deployment $M_1 = 15 (1 - \omega/10)$, $M_2 = 24(1 - \omega/0.1)$. The analysis shows that with the system of eight beams the central cylinder angular velocity would decrease to zero assuming that seven meters of the radial structure have been released, the total length of the beam being 13.74 m, that is, the deployment would come to a stop. For this reason, the 12-beam deployment scheme is proposed. Dynamic parameters for the first step with $n = 12$ are given in Fig. 3.18.

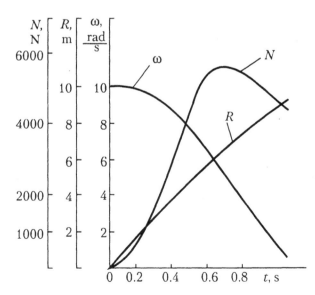

3.18. Uncontrolled deployment dynamic parameters (unfolding from the central cylinder) at step 1. $\omega$ — angular velocity of the central cylinder rotation; $N$ — sum tension in the root cross-section; $R$ — structure radius

Once the beams are deployed and the remainder of the reflector has yet to be deployed, oscillations of the beams of the 1.5 amplitude would be observed. No further deployment is suggested for 200 seconds in which, according to the calculations, the oscillations would be damped to +0.38 rad. If the deployment was not initiated at this time, the ultimate stress would be exceeded thereafter as a consequence of an increase of the central cylinder angular velocity of rotation.

At the second step of the reflector deployment (a controlled deployment of the rest of the reflector) the initial value of the reflector tilt angle is taken to be +0.38 rad. The calculations have shown that the deployment is oscillatory in nature at all velocities above 1 cm/s. With the deployment velocity of 1 cm/s, the oscillations were damping and the second-step deployment had been smoothly accomplished in 3500 seconds. Dynamic parameters of the second step deployment process are given in Fig. 3.19. This mode can not be recommended because of the unacceptably large amplitude of oscillations at the initial step.

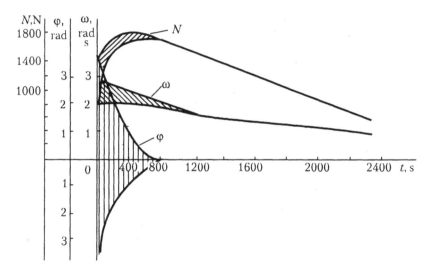

Fig. 3.19. Dynamic parameters of the solid reflector deployment at step 2 (controlled deployment of the reflector remainder). $\varphi$ — angle of deflection from the radial direction; $\omega$ — angular velocity of the central cylinder rotation; $N$ — total tension in the root cross-section

## 3.5. Mathematical Model of Tether System Deployment

The procedure used to configure the tether system (antenna, flywheel) is shown in Fig. 1.3. The central cylinder accommodating spools on which the system is reeled up is set in rotation, whereupon, by the action of centrifugal forces, first the circumferential and then radial tethers are fed out in steps. It thus appears that the system deployment is simulated in two steps: first, the deployment dynamics is simulated for the tethers which will subsequently shape the periphery, then the deployment step is simulated for the radial tethers in a manner when the circumferential tether is replaced with a broken one consisting of two straight members of constant mass

$$m = \mu \frac{\pi}{n} r_k .$$

Step I: Simulation of the deployment process for circumferential tethers (arranged in parallel pairs with free ends).

Let us write the principal equation of dynamics for the tether element:

$$\mu \, dl \, \ddot{\overline{R}} = d\overline{N} ,$$

where $\mu$ — the tether linear density.

Using (3.9) and projecting on the axis of baseline $e^{(2)}$, integrating along the entire length and multiplying by $2n$ we obtain:

$$N = 2 \mu L \left[ r_0 ( \omega^2 \cos \varphi - \dot{\omega} \sin \varphi ) - \ddot{L} + 0.5 L ( \omega + \dot{\varphi} )^2 \right] \qquad (3.38)$$

$$r_0 ( \dot{\omega} \cos \varphi + \omega^2 \sin \varphi ) + 2 ( \omega + \dot{\varphi} ) \dot{L} + 0.5 ( \dot{\omega} + \ddot{\varphi} ) L = 0 \qquad (3.39)$$

Hence, we have combined equations (3.14, 3.38, 3.39). The initial conditions are:

$$\varphi ( 0 ) = \dot{\varphi} ( 0 ) = 0 , \omega ( 0 ) = \omega_0 .$$

Step II: Simulation of the deployment process for radial tethers. Initial values of variables at the second step are their final values determined at the first step.

The equation of motion for the central cylinder is similar to (3.14).

Similarly to (3.38, 3.39), the expression for the radial tether is:

$$N_1 - N_2 = \mu L \left[ r_0 ( \omega^2 \cos \varphi - \dot{\omega} \sin \varphi ) - \ddot{L} + 0.5 L ( \omega + \dot{\varphi} )^2 \right] \qquad (3.40)$$

where $N_2$ — tension of the radial tether in the point of coupling with the circumferential tether.

$$2 n \mu L ( r_0 ( \dot{\omega} \cos \varphi + \omega^2 \sin \varphi ) + 2 ( \omega + \dot{\varphi} ) \dot{L} +$$

$$+ 0.5 ( \dot{\omega} + \ddot{\varphi} ) L ) = F_z \cos \alpha \qquad (3.41)$$

where $F$ — force directed normally to the radial tether and equal to the force applied to the radial tether by the circumferential one.

$$F_z = 2 n F .$$

From geometrical considerations (Figs. 3.20, 3.21) we have:

$$\cos \alpha = \frac{L + r_0 \cos \varphi}{R},$$ (3.42)

$$R^2 = L^2 + r_0^2 + 2 L r_0 \cos \varphi$$ (3.43)

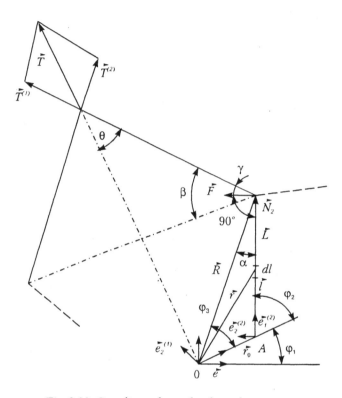

Fig. 3.20. Coordinate frame for the tether system

Hence, the theorem of angular momentum variation for the circumferential tether in the projection on the axis of rotation will be written:

$$\frac{d}{dt}\left[ I_{circ} (\omega + \dot{\varphi}_3) \right] = - R F_z,$$ (3.44)

$$\sin \varphi_3 = \frac{L \sin \varphi}{R};$$ (3.45)

$$I_{circ} = \frac{2}{3} n \mu R_{kn} ( R_{ki}^2 + R_{ki} L + L^2 ),\ R_{kn} = \pi R_k / n;$$

$$R_{ki} = R_{kn} \cos \theta + R_{kn} \sin \theta \cdot \text{ctg} \frac{360°}{2 n}$$

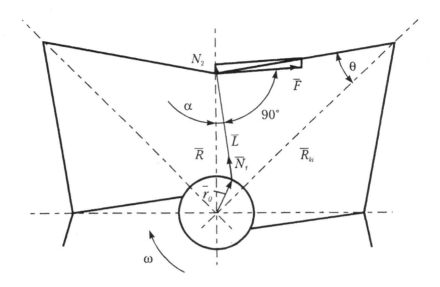

Fig. 3.21. Step 2 of the tether system deployment

Equations (3.14, 3.40, 3.41) contain 6 unknowns, $N_1$, $N_2$, $\omega$, $\varphi$, $F_y$, $\alpha$.
Three coupling equations (3.42, 3.43, 3.45) produce the seventh unknown, $R$.
Equation (3.44) adds a new unknown, $\varphi_3$.
To close the system, the equation of force $N_2$ being the projection of tensile force $T^{(1)}$ in two adjacent members to direction $L$. The resulting tensile force $\overline{T} = \overline{T}^{(1)} + \overline{T}^{(2)}$ (Figs. 3.20, 3.21) is:

$$T = \frac{2\,\mu\,R_{kn}}{(R_{ki} - L)} \int_L^{R_{ki}} \left[ r_0(\omega_1{}^2\cos\varphi_2 - \dot{\omega}_1\sin\varphi_2) - \ddot{L} + L(\omega_1 + \dot{\varphi}_2)^2 \right] dL =$$

$$= 2\,\mu\,R_k \left[ r_0(\omega_1{}^2\cos\varphi_2 - \dot{\omega}_1\sin\varphi_2) - \ddot{L} + (\omega_1 + \dot{\varphi}_2)^2 (R_k + L)\,0.5 \right]$$

$$N_2 = \frac{T\sin\gamma}{2\cos\theta}$$

The radial tether stress in the point of coupling with the circumferential tether is obtained from the equation:

$$N_2 = 2\,\mu\,R_k \Big[ r_0(\omega_1^2\cos\varphi_2 - \dot{\omega}_1\sin\varphi_2) - \ddot{L} +$$

$$+ (\omega_1 + \dot{\varphi}_2)^2 (R_k + L)\,0.5 \Big]\sin\gamma\,/(2\cos\theta), \qquad\qquad (3.46)$$

where

$$\sin\theta = L\sin\frac{360°}{2\,n}/R_{kn}; \quad \gamma = \frac{-360°}{2\,n} - \theta + 90°.$$

The possibility of controlling a mechanical system consisting of a central cylinder, flexible links, and a deployment mechanism having different characteristics has been explored by an example of a tethered antenna of up to 500 m in diameter.

Through analysing the numerical calculation results the following conclusion has been made: at the first step (simulated deployment of circumferential tethers) the system remains relatively stable, that is, at the deployment velocity of 0.1 m/s and with characteristics of two electric motors of the drive being $M_1=1.5(1-\omega/8)$, $M_2=24(1-\omega/0.75)$, $\mu = 2.6 \cdot 10^{-3}$ kg/m, $R_0 = 1.1$ m a smooth deployment of the circumferential tethers will be observed (Fig. 3.22), $\varphi_{max} = -0.278$ rad., $N_{max}=168$ N, $\omega_{min} = 0.4$ rad/s. The circumferential tethers will be deployed in 38 min. However, when the deployment is commenced at an angular velocity of 0.4 rad/s at the moment corresponding to this point, the system will be get into a tangle (with angle $\varphi$ exceeding $\pi/2$ ) with the system final radius over 60 m. The objective was brought of the selection of deployment parameters for the antenna of $R = 270$ m radius. Factors effecting the system stability have been studied. These are: an increased radius of the central cylinder radius $(R_0)$, angular velocity $\omega$ (for $R_k = 270$ m, $\omega_{max} = 0.7$ rad/s) increased to the maximum possible values (proceeding from the material strength), the moment imparted from the electric motor decreased to values close to zero, an increased number of sectors. Consequently, increasing the central cylinder radius to $R_0 = 3$ m will increase the structure final radius to $R_k = 150$ m with $R_0 = 5$ m, $R_k = 190$ m; with $R_0 = 10$ m, $R_k = 230$ m, respectively. The central cylinder mass could be increased through the accommodation within this mass of a radio-sig-

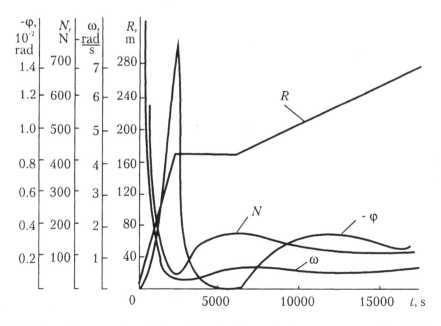

Fig. 3.22. Dynamic parameters of the tethered antenna deployment. $\varphi$ — angle of deflection from the radial direction ($10^{-2}$ rad); $\omega$ — angular velocity of the central cylinder rotation (rad/s); $N$ — total tension in the root cross-section ($N$); $R$ — structure radius (m)

nal generator supplying RF power to the antenna. The estimation results show that increasing the central cylinder mass to 800 kg allows, at $R_0$ = 3 m, a smooth deployment of the structure with a final radius of 270 m.

One of the antenna deployment scenarios with $R_0$ = 10 m, $R_k$ = 230 m involves, at the second step, an increase of the initial angular velocity and a decrease of the initial external moment to $\omega_{init}$ =0.7 rad/s, $M_{init}$ =0.1H·m, respectively (Fig. 3.22). While unreeling the circumferential tether, $\omega$ is diminished to 0.4 rad/s. Prior to initiate the deployment of radial tethers, the structure has to be run to angular velocity of $\omega$ = 0.7 rad/s. It is seen from the calculations that the procedure requires one hour. From there, the deployment of radial tethers is commenced. The following dynamic parameters for the second-step deployment have been obtained (Fig. 3.22): $\varphi_{max}$ = 0.3 · $10^{-2}$rad, $\omega_{min}$ = 0.62 rad/s, $N_{max}$ = 180 N.

In performing design studies on the system, to ensure its electrical performance and improve its arrangement on a carrier, the polyimide-insulated copper tether 0.1 mm thick with the distributed mass of $\mu$ =23 $10^{-3}$ kg/m was selected in the capacity of a conductive tether. The polyimide insulation was selected against other candidates due to its thermal resistance, $T_w$ < 450°C, with a high value of penetration stress retained, $U_{penetr}$ = 2000V, as well as the possibility to practically ignore the insulation mass because of its small thickness in calculating the antenna dynamics, as the insulation mass will be of the order of 3% from the mass of copper. Calculations involving new values of $\mu$ have been repeated with all major tendencies preserved.

The antenna angular rotation velocity is selected on the basis that the residual strain stresses in the tether materials are exceeded due to the fact that, upon deployment, the tether will remain coiled and, to make it straightened, forces have to be applied experimentally estimated as $\sigma$ = $10^6$ N/m². Based on this consideration, the final rotation velocity of the antenna is $\omega$ = 0.1 rad/s. Estimations of other orbital impacts resulted in lower values. To increase the system initial radius, $R_0$, is a rather complicated procedure from the engineering viewpoint. It would be reasonable to increase a number of antenna sectors to make the deployment stable. The estimations suggested that by increasing a number of sectors from 4 to 8 the system stability would be efficiently enhanced and the additional radial links could be non-conducting.

## 3.6. Control of Counter-Rotating Systems

In compensating the reflector angular momentum at the expense of a tethered counter-rotation flywheel (or a counter-rotation reflector), the space vehicle angular velocity is required to be zero throughout the deployment. This requirement should be met when moments imparted from electric motors are equal and should be provided through the control system with a feedback for indications of angular velocity sensors located on three axis of the vehicle.

An equation of the vehicle motion relative to the longitudinal axis:

$$J_{vch} \, \dot{\omega}_{vch} = M_1 - M_2,$$

where   $J_{vch}$ — inertia moment of the vehicle relative to that axis;
    $\omega$ — angular velocity of the vehicle;
    $M_1$ — external moment applied to the reflector;
    $M_2$ — external moment imparted to the flywheel.
To make $\dot{\omega}$ equal to zero (in this case $\omega = const = \omega \, ( \, 0 \, ) = 0$ ), it is required that $M_1 = M_2$.

## 3.7. Repackaging Control

### 3.7.1. Repackaging Process Features

For a number of applied tasks a necessity was brought of how to repackage flexible elements of a system. For example, how to repackage and deliver a sensitive surface of the space debris recording system to the Earth; how to control the tethered antenna loop to adjust it to a required low frequency, etc.

This Chapter describes how to repackage the transformable systems (film and tether) to turn them to the original or other patterns using the scientific-engineering knowledge.

The repackaging process differs from the deployment process in several essential points. First, it is extremely difficult and actually unfeasible to repackage complicated multi-link film patterns originally packaged as being fabricated. The possible robotic film stacking mechanism appears to be much more complicated comparing to the film deployment mechanism itself. The most efficient repackaging process is offered by ribbon and tether structures. For the tether structures a tether stacking mechanism is required.

The rotating system repackaging process is peculiar in that the angular momentum accumulated during deployment should be taken off from the system. Principally, three ways of compensating the angular momentum are possible: by consuming a working medium of the vehicle reaction control system; by "transferring" to a system being deployed in parallel; by repackaging a compensating system when the system has been originally based on two counter-rotating systems.

It thus appears that with large structures and the large angular momentum an inexperience of using the reaction control system is evident. Use of two counter-rotating systems on one vehicle gives the ability to employ both the "transferring" mode and the full compensation mode.

### 3.7.2. Mathematical Model of Repackaging Process

The repackaging dynamics can be treated by using the mathematical simulation similar to that employed for solving the deployment tasks with the peculiarity applying that the law of change of mass is changed in the film sheet task and a control braking moment, rather than a moving moment of rotation is sought for. Accordingly, the sign is changing versus the deployment velocity.

The objective is to establish the control laws for the braking moment and repackaging velocity giving the ability to fold back the film structure with no intolerable oscillations and entanglements and in a required timeframe. For this, the stress applied across the structure root cross-section should not exceed a tolerable value for the given material.

The mechanical system (equations 3.14, 3.38, 3.39) consisting of a flexible element (cable or ribbon), a central cylinder, and a rotation drive which performance depends on stability parameters $\gamma = L\omega/(2V)$ ($V$ — retraction velocity) has been investigated for the repackaging process stability.

With the drive moment equal to zero the structure angular velocity of rotation increases in accordance with the law of momentum accumulation by the system. A stable repackaging process is observed with no oscillations, at a small angle of deviation and a large value of stability parameter $\gamma = L\omega/(2V_{repack}) = 100...500$, where $V_{repack}$ — repackaging velocity. However, in this case increasing the system rotation angular velocity will lead to exceeding the ultimate stress and to the structure failure. Consequently, a negative momentum has to be applied to the system (a moment of momentum has to be taken off). The objective is posed to explore possible laws of controlling the braking moment and the repackaging velocity responsible for the system stability. When the drive characteristics are $M = -M_0(1 - \omega/\omega_0)$ or $M$=const, the central cylinder will be initially set in rotation at a negative angular velocity and the angle of deflection from the radial direction $\pi/2$ will be exceeded (the system will be get in a tangle) that is technically inadmissible. It has been found out that boundaries of zone of stability for the imparted momentum are going to be $M = 0.1\omega - -0.3\omega$. Modes with different repackaging velocities are under consideration. Fig. 3.23 shows the dependence of the structure deflection angle on dimensionless parameter $\gamma = R\omega/2V_{repack}$, where $1 - M$=-0.1$\omega$, $V_{repack} = 0.5$ m/s; $2 - M$=-0.1$\omega$, $V_{repack} = 0.5$ m/s; $3 - M = -0.3$, $V_{repack} = 0.2$ m/s. It is seen from the Figure that in the first case a smooth repackaging takes place ($\varphi < 0.2$ rad), however, the angular velocity of rotation will exceed the value acceptable for the given structures reasoning from strength considerations. In the second and the third cases angle $\varphi$ will exceed 1 rad at the process completion. Modes with the acceptable angular velocity of rotation and angle of deflection from the radial direction are intermediate. Besides, one of possible repackaging options involves an electric motor shutdown prior to the process completion ($M_{t=0.8}=0$). Fig. 3.24 shows angle ($R = R/R_k$) and angle of deflection $\varphi$ versus time ($t = t/t_k$) at $M$=-0.1$\omega$, $V_{repack} = 0.05$, where $1$ — time dependence of the structure radius, $2$ — time dependence of the angle of deflection, $3$ — variation zone of the angle of deflection when the motor is shut down at $T$=0.8, $4$ — zone in which the angle of deflection is varying when the motor is shut down at $T$=0.85. As evidenced from the diagrams, oscillations are observed in the given mode, however, the amplitude is acceptable. In this way the repackaging process will go at an admissible angular velocity of rotation.

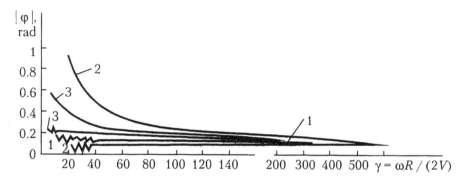

Fig. 3.23. Angle φ versus dimensionless parameter gamma during repackaging where 1—
$M = 0.1\omega$, $V_{\text{repack}}=0.5$ m/s; 2 — $M = 0.1\omega$, $V_{\text{repack}}=0.05$ m/s; 2 — M = 0.3ω, $V_{\text{repack}} = 0.2$ m/s

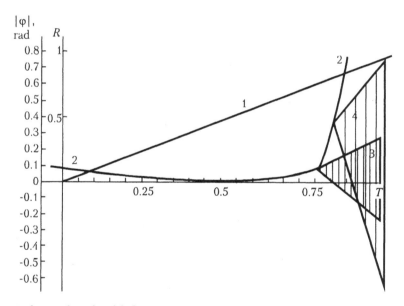

Fig. 3.24. Radius and angle of deflection versus time. 1 — structure radius; 2 — angle of structure deflection; 3 — angle φ variation zone at the motor shutdown with $T = 0.8$; 4 — angle φ variation zone at the motor shutdown with $T = 0.85$

This analysis suggests the conclusion stating that the procedure of repackaging the dynamic structures could be controlled by using the d.c. motor operated in a generator mode to provide a braking moment.

It must be emphasised that the repackaging control process, as mentioned above, could be provided by using the d.c. motor operated in a driving mode. The d.c. motor is highly reliable, compatible with other elements of the space vehicle control system, convenient and easy for ground processing. The given systems are unique in the possibility of using the d.c. motors to control both the deployment and repackaging of the dynamic system.


## 3.8. Impacts of Longitudinal, Lateral, Torsional Oscillations and Space Environment Factors

The aforesaid analysis has shown that the repackaging-capable, the base of the deployable, flexible system would suffer from periodic perturbation forces associated with variations of angle φ relative to the central cylinder and resulted in periodical changes of tension in the root cross-section of the element being repacked. The probability exists that the constrained longitudinal, lateral, and torsional oscillations are excited in the system flexible element leading to the undesirable response of the system, as a whole (flexible element, central cylinder, motor performance). Such a phenomenon could be expected when the tension variation frequency in the flexible element root cross-section would coincide with the frequency of free longitudinal, lateral, and torsional oscillations.

Oscillations in the tense, flexible element are described in the known wave equation; propagation velocity $V$ of longitudinal oscillations is given by:

$$V = ( E / \rho )^{1/2} ,$$

where $E$— modulus of elasticity of the material;
   $\rho$— material density.
For film materials $E = 10^9 \, \text{N/m}^2$, $\rho = 1.4 \, \text{kg/cm}^3$ and $V$ is of the order of 1000 m/s.

A period in which oscillations propagate along the element is equal to a doubled period of wave propagation ($t = 2L/V$, where L is the element length). The aforesaid assessment shows that the frequency of free longitudinal oscillations is many orders of magnitudes higher than the constrained frequency and any resonance is impossible.

The propagation velocity of lateral oscillations is given by

$$V = ( \sigma / \rho )^{1/2} ,$$

where $\sigma$— stress of the tense, flexible element;
   $\rho$— material density.
A tension in the element expanded by centrifugal forces is alternatively variable for film materials within a range from $\sigma = 10^7 \, \text{N/m}^2$ to $\sigma = 0$. Along the greater part of the element the material stress is $\sigma = 10^6 \, \text{N/m}^2$. Consequently, the frequency of free lateral oscillations will be an order of magnitude lower against the longitudinal

oscillation frequency and, however, will be also well above the constrained oscillation frequency.

In principle, the torsional oscillations frequency is of the order of that of the lateral oscillations. However, in certain structures under consideration no regular perturbations occur to support such perturbations.

Analysis of the in-orbit impacts on large structures suggests that the major in-orbit disturbing impacts are: gravitational, aerodynamic effects, and light pressure. For systems expanded to dimensions of up to 500 m by centrifugal forces the impact from the aforesaid factors is negligible due to the fact that the structure working tension from centrifugal forces is essentially surpassing the in-orbit forces.

## 3.9. Simulation of Loads Applied to Electric Drives During Ground Verification

A major peculiarity of developing the space-based, large ($D$=100, 200 m) structures is their impossible full-scale ground verification owing to the gravitation effect and the lack of vacuum chambers of such a large size.

During the in-orbit experiment (the Znamya-2 experiment) successfully completed on February 4, 1993 on the Progress vehicle the $D$=20 structureless film reflector was the first-ever complicated tethered film structure of 32 tethers and 8 film sectors to be deployed in space.

Besides the routine bench and qualification tests conducted during the ground verification of space hardware items, the dynamic simulation techniques have been developed and implemented. The objective was to generate, by means of tethered weights, loads acting on electric drives of the reflector deployment mechanism in a vacuum chamber of limited dimensions and similar to forces generated by the film reflector throughout the in-orbit deployment steps and further operation. Additionally, in the same vacuum chamber the mechanics of unreeling the tethered film structure from spools has been specially verified at the initial, intermediate, and final steps of the reflector deployment.

The dynamic simulation technique is based on the solution of the second order differential equations describing the deployment dynamics of a solid film sheet and a tethered weight system with the weight masses selected to simulate, at the highest accuracy, the principal dynamic parameters of the deployment process, such as the variable law of drive loading, angle of deflection from the radial direction, angular velocity of rotation.

Mass of the simulating weight can be calculated by equalising the coefficients at the final terms of in the left part of equations (3.11) and (3.15) and assuming $L = L_k$ we have:

$$m = \frac{2\pi\beta \cdot L_{k1}^3}{3nL_{k2}},$$

where β — film surface density;

      $n$ — number of reflector sectors (weights);

      $L_{k1}$ — final radius of the reflector;

      $L_{k2}$ — final radius of the tether and weight.

It is worthy to note that the detailed simulation does not appear to be accurate and rigorous likely because the solution of the equations is weakly dependent on a specified mass of the flexible system. However, this simulation appears to be rather useful due to the fact that in actual practice a functional margin is required which sufficiently surpasses the solution accuracy of the equations.

As follows from the numerical calculations, the dynamic impact to the electric drive from the film sheet of 4.5 kg mass and 10 m final radius being deployed could be simulated by the 16 kg tethered weight released at the radius of 1.7 m. Here, the deployment time is reduced from 200 s to 30 s. Dynamic parameters of the reflector being deployed and the system of tethered weights simulating the reflector are given in Fig. 3.25 for the electric drive characteristic selected for the experiment. To test the deployment mechanism the vacuum chamber 10 m high and 5.5 m in diameter designed at RSC Energia was employed.

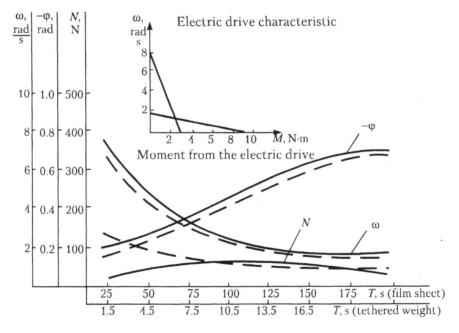

Fig. 3.25. Simulation of dynamic characteristics of the reflector deployed in the space-based Znamya-2 experiment by means of tethered weights

The given technique should be used for testing large structures built by centrifugal forces.

It should be also noted that, as a whole, this approach has to be applied to other systems deployed in space (tether, framed, etc.). Analyses of a large number of experiments suggests that tests in the open area would be insufficient, and hence it appears that specific features of the deployment dynamics should be simulated in vacuum chambers during ground tests. While performing the ground simulation the loads should be at most similar to those acting in the real scenario and have a margin to assure the system actuation reliability.

## 3.10. Conclusions

The techniques proposed in this Chapter for deploying space structures by centrifugal forces and their repackaging are based on the aforesaid limitations (3.5, 3.6) establishing the predominance of exactly centrifugal forces, their stabilising and shaping action in the processes under consideration. This allowed to make the process less complicated by assuming the radial rectilinearity of the sheet fold (strip, tether), no dissipation and no need to consider a specific configuration of a fold and its spacial changes in time.

The validity of this approach has been proved by comparing the estimated data obtained under the aforesaid techniques with results of measuring the similar characteristics during the ground and in-orbit experiments. (See Chapter 7).

The developed methodology includes the selection of folding patterns for film structures and tethered systems, equipment providing the rotation, deployment and repackaging, as well as suggests methods of mathematical and dynamic simulation for ground verification. The techniques can be applied to a wide spectrum of practical tasks where spacecraft structures are not required to be deployed at high velocities.

# CHAPTER IV

# IN-ORBIT RE-ORIENTATION OF STRUCTURES EXPANDED BY CENTRIFUGAL FORCES

## 4.1. General Statements

In performing attitude control of rotating film reflectors and tethered systems the flexible systems are suffering from gyroscopic forces causing their surface shape to change. This change is manifested by occurring on the rotating structure of a skew-symmetric travelling wave reversibly directed relative to the system rotation. The wave amplitude is maximum at the structure periphery,

$$A \approx \frac{2 \Omega R_k}{\omega R_0},$$

and depends on the structure geometry (ratios $R_k/R_0$, where $R_k$ — the system outer radius, $R_0$ — radius of the guiding centre through which a moment of rotation is imparted to the structure) and the $\Omega/\omega$ ratio of the system rotation in space, $\Omega$, to a velocity of its rotation around the central axis, $\omega$. (See below). In a system with two counter-rotating structures the amplitude value determines the possibility of entanglement between the structures and also establishes a length of a baseline between the rotating structures (considering an angle of deflection or, as is customary to call, an angle of actuation). Surface accuracy limitations associated with the surface deformation caused by gyroscopic forces are becoming of a prime concern for rotating systems where increased requirements are imposed on the surface accuracy and orientation (e.g. reflectors for illuminating the Earth's regions with reflected sunlight). In principle, two ways exist to decrease the undesirable amplitude of travelling wave. First, geometrical parameters of rotating structures and relative rotation velocities are selected so as to ensure a small amplitude. It could be possible by using, instead of one large reflector, a large number of small reflectors combined into one system by means of tethers. As the angular velocity of rotation depends on the material strength and is inversely related to the system radius,

$$\omega \approx \frac{I}{R} \sqrt{\sigma / \rho} ,$$

then, for the film reflectors with $R = 15-30$ m the angular velocity of rotation can be relatively high thereby decreasing the amplitude of oscillations.

The second way is to provide active or passive damping of oscillations. That the second way is feasible has been proved by experiments on the rotating sheet oscillation in a vacuum chamber Y-22 at TSNIIMASH. Unfortunately, this task has not been solved until now and is still remaining a problem. An active damping of oscillations has been rather successfully attempted during the space-based Znamya-2 experiment, however, to obtain a high accuracy of surface would require further experimental and theoretical research.

## 4.2. Mathematical Model of Tethered Weight Rotation

Since the reflector element can be taken as a localised mass of a tethered weight, commence to treat a rotating structure turn in space just from this model which, in the same manner as in the task where the rotating structure is fed out from the folded pattern, is physically and mathematically demonstrative while possessing many properties of systems with distributed parameters.

Fig. 4.1 shows mass point m connected to weightless, unstretchable tether $R_2$ long attached to a guiding, rigid central insert $R_0$. The mass is rotating about the central axis $0_1 0$ at angular velocity $\omega$ and spinning at angular velocity $\Omega$ about the axis perpendicular to $0_1 0_2$ and lying within the drawing plane. As this takes place, in a position shown in Fig. 4.1 the mass is exerted by centrifugal force $F_{cf} = m \omega^2 R_1$. The Coriolis force $F_{cor} = 2 \omega \Omega R_1 m$ and the mass is deflected at angle $\varphi$ within the drawing plane.

In the coordinate system rotating at velocity $\omega$ and referenced to the central cylinder the weight equation of motion may be written as:

$$m R_2^2 \ddot{\varphi} + m \omega^2 R_1 R_2 \sin \varphi = 0 \qquad (4.1)$$

and represents an equality of moments relative to the tether attachment point on the guiding central insert.

As far as small oscillations ($\sin \varphi \approx \varphi$) are under consideration, the equation will take the form:

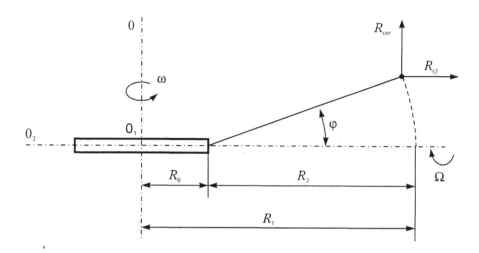

Fig. 4.1. Tethered mass point rotation

$$\ddot{\varphi} + p^2 \varphi = 0 \tag{4.2},$$

where

$$p = \omega \sqrt{R_1 / R_2} \tag{4.3}$$

is the system natural frequency.

Hence, equation (4.2) will be solved as:

$$\varphi = A \sin\left[ \omega \sqrt{R_1 / R_2}\, t + \theta \right] \tag{4.4}$$

Equation (4.2) demonstrates that, with the central insert $R_1 > R_2$ the system natural frequency is always higher than rotation angular velocity $\omega$.

Within the rotating coordinate system referenced to the central rigid insert the Coriolis force acting on the mass is serving to provide a harmonic effect at rotation angular velocity $\omega$, viz:

$$F_{cor} = 2\,\omega\,\Omega\,R_1\,m \sin(\theta + \omega t) \tag{4.5}$$

In this case the steady-state oscillation amplitude is given by the equation

$$\varphi = \frac{\varphi_{st}}{\sqrt{(1 - \dfrac{\omega^2}{p^2})^2 + \dfrac{4\,\omega^2\,n^2}{p^4}}} \tag{4.6}$$

where $\varphi_{st} = 2\dfrac{\Omega}{\omega}$ — angle of tether rotation at an applied static load;

$\qquad\qquad n$ — coefficient describing the system dissipation properties.

Substituting (4.3) in (4.6) with $n=0$ the amplitude of oscillations will be defined by

$$\varphi = \frac{\varphi_{st}}{1 - R_2/R_1} \qquad\qquad (4.7)$$

or, considering that $R_1 - R_2 = R_0$,

$$\varphi = \frac{2\,\Omega}{\omega}\frac{R_1}{R_0} \qquad\qquad (4.8)$$

With $n=0$ equation (4.6) also suggests that

$$\varphi = \frac{2\,\Omega\,p^2}{\omega\,(p-\omega)(p+\omega)} \qquad\qquad (4.9)$$

With $R_1 > R_0\ p \approx \omega$ , equation (4.9) may be also given as

$$\varphi = \frac{\Omega}{p-\omega} \qquad\qquad (4.10)$$

An outside observer located within the fixed coordinate system would view the travelling wave on the surface of the reflector rotating at frequency $p - \omega$ with the wave rotating against $\omega$ , as $p > \omega$.

It is notable that in the space-based Znamya-2 experiment the frequency difference, $p - \omega$, was measured with the angular velocity sensors for the period of the travelling wave, $T=40$ s, the rotation velocity was set as $\Omega = 0.2$ deg/s. In this way the angular amplitude of oscillations after (4.10) will be expressed by

$$\varphi = \frac{0.2 \cdot 40}{360} = 2 \cdot 10^{-2}\ \text{rad} = 1.2°,$$

The obtained results were in a fair agreement with other calculations.

## 4.3. Mathematical Model of Rotating Tether with Attached Distributed Mass and Weight at the End

Consider the case where a tether of mass $m_0$ is carrying a weight of mass $m$.

By substituting the corresponding integrals for the tether length in equation (4.1) we obtain

$$\int_{R_2}^{R_2} \mu r^2 dr + m R_2^2 = (\frac{m_0}{3} + m) R_2^2$$

$$\int_0^{R_2} \mu r dr + m R_2 = (\frac{m_0}{2} + m) R_2$$

(4.11)

where $\mu$ — distributed mass of the tether;
$\mu R_2 = m_0$ — total mass of the tether.
Whence it follows that the system natural frequency is

$$p = \omega \sqrt{(m_0/2 + m) R / (m_0/3 + m) R_2}$$

(4.12)

Equation (4.12) suggests that with a low mass of the end weight, $m_0 > m$, the frequency is higher than with the only weight available (4.3) and, hence, the amplitude given in (4.6) would be lower.

It thus appears that, with the periphery mass of the tether or localised masses on the reflector, an undesirable increase of the sheet edge deflection would be observed in the process of the sheet attitude control.

It must be emphasised that equation (4.8) derived for the tethered weight system allows to obtain a maximum possible value of the angular amplitude. In systems with the peripheral linkage (tether ring or split sheet) a consideration must be given to effects from adjacent areas which prevent the system from deflecting out of the plane of rotation under the Coriolis forces and lead to a decrease of the static amplitude value ($\varphi_{st}$) which could be estimated from the expression close to a catenary at small deflections, viz

$$N_\theta = \frac{q_u R_1^2}{8 f}$$

(4.13)

where $f$ — deflection (see below).
For a tangential force write similar to a rotating ring, viz

$$N_\theta = S \rho \omega^2 R_1^2,$$

where $S$ — tether cross-section;
$\rho$ — material density.
Suppose the distributed load $q_u$ is a load generated by the Coriolis forces and referred the chord length equal to the system radius, viz:

$$q_u = 2 \omega \Omega S \rho R_1$$

Hence, for the static angular amplitude we obtain:

$$\frac{f}{R_1} = \frac{2\,\Omega}{8\,\omega} = \frac{\varphi_{sl}}{8} \tag{4.14}$$

It thus appears that the peripheral linkage, comparing to weights not linked in between (4.8), decreases the static amplitude almost by an order of magnitude.

## 4.4. Mathematical Model of Rotating Solid Sheet

A system of differential equations describing how the rotating membrane configuration is changed in time and defining a stressed-strained state of the object in the process of complicated rotation is rather inconvenient.

It was evident from the calculation results that, with the specified geometrical and physical parameters of the structure and in the motion modes under study, additional migrations $\Delta u$ and $\Delta v$ of the membrane points within its plane caused by the membrane axis of·rotation are five orders of magnitude less than migrations $\Delta w$ out of the membrane plane. Hence, the solution is reduced to one differential equation, viz

$$N_r^* \frac{\partial^2 w}{\partial R^2} + \frac{N_\theta^*}{r}(\frac{\partial w}{\partial r} + \frac{1}{r}\frac{\partial^2 w}{\partial \theta^2}) - \rho\,h\,\omega^2\,r\frac{\partial w}{\partial r} - \rho\,h\frac{\partial^2 w}{\partial t^2} - c\frac{\partial w}{\partial t} =$$
$$= \rho\,h\,r\left[\,2\,\omega\,\Omega\,\sin(\omega\,t+\theta) - \frac{\partial\Omega}{\partial t}\cos(\omega\,t+\theta)\,\right] \tag{4.15}$$

Here, $r, \theta$ — radial and tangential factors of the membrane points; $\rho, h$ — density and thickness of the membrane; $R_0$ — radius of the central rigid insert; $N_r^*$ and $N_\theta^*$ — radial and tangential forces generated in the membrane at its free rotation.

Numerical calculations based on the Haubolt method appeared to be in a fair agreement with the aforesaid assessments and results obtained from the Znamya-2 experiment.

Research has been conducted in oscillations of the membrane with the following geometrical and physical parameters: $R = 10$ m, $R_0 = 0.1$ m, 0.2 m, 0.6 m, 1.0 m; $h = 10^5$ m; $E = 3 \cdot 10^9$ N/m²; $\nu = 0.3$; $\rho = 1.3 \cdot 10^3$ kg/m³.

Fig. 4.2 demonstrates the case when the membrane axis angular rotation velocity is abruptly increasing and the axis is further rotating at a constant velocity. The oscillation amplitude is drastically suffering from dissipation coefficient $C$ and ratio $R_0/R$ between the rigid insert radius and the membrane radius. Commencing from $R_0/R = 0.1$, the growth of oscillations is terminated at a certain moment of time. Fig. 4.3 illustrates the edge deflection with similar parameters $\omega$ and $\Omega$ for different systems at $R = 100$ and $R_0 = 20$ m.

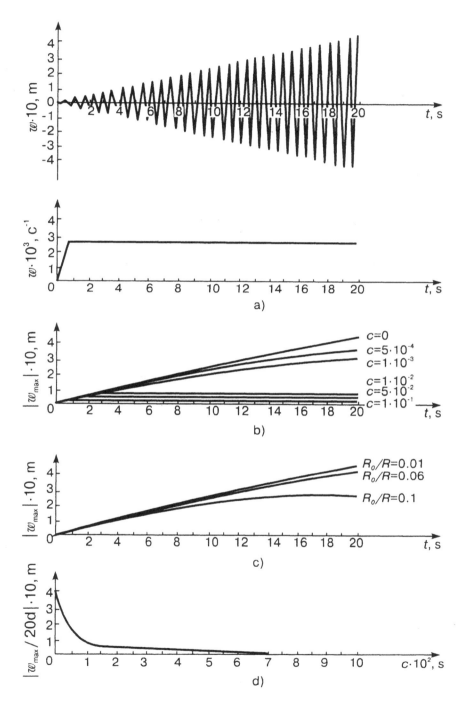

Fig. 4.2. Membrane oscillations at the abrupt increase of the axis rotation angular velocity and further rotation at the constant angular velocity

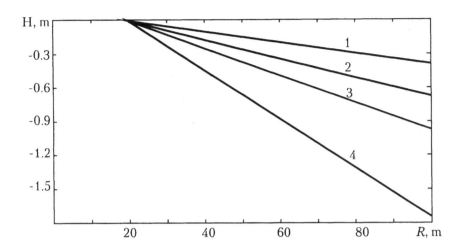

Fig. 4.3. Amplitudes of deflection out of the plane of rotation calculated for different systems.
1 — tether ring; 2 — split sheet combined around the periphery; 3 — tethered weight; 4 —
solid sheet; $\omega = 0.5$ rad/s; $\Omega = 5 \cdot 10^{-4}$ rad/s

# CHAPTER V

# STRUCTURES, MATERIALS AND PROCESSES USED TO FABRICATE FILM REFLECTORS, STRENGTH AND REFLECTANCE

## 5.1. Film Structure Operation Environment

For purpose of solving various applied tasks and performing space-based experiments, low orbits of an altitude under 400 km and high orbits from 1600 km to the geostationary orbit are being considered. Interplanetary travels on solar sailing vehicles make possible the exploration of deep space mainly depending on the distance from the Sun. The major factor limiting the film material service life in low orbits is its reaction to the residual atmosphere components, predominantly, oxygen atoms and ions. The reflection performance of coatings depends on a relative deformation of a substrate and comes to a maximum at values of relative deformation close to a yield stress of the material. At this deformation the surface inaccuracies associated with the fabrication tolerances are removed (that is during operation the reflector surface is in the two-dimensional stressed state). Depending on the distance from the Sun and the Earth, as well as on the orientation in space the direct and reflected sunlight fluxes are falling upon the film structure in a wide range of wavelengths thereby heating the film and causing its structural changes. According to a combination of the solar radiation absorptivity value, $A_s$ and the surface blackness level in cases of the one-side or two-side metal coating, the surface temperatures from $T = 270°C$ to minus values could be obtained in the near-Earth orbits. As this takes place, the temperature is varying in different portions of the orbit as the angular position is different relative to the Sun due to the reflector rotation. Depending on the orbit altitude, the reflector material, besides the solar radiation, is subjected to:

• galactic cosmic ray charged particles,
• Van Allen belt protons and electrons (at an altitude in excess of 1000 km).

The galactic cosmic ray flux consists of protons and heavy nuclei of superhigh energies arriving from remote fields of the Galaxy. The flux intensity is 2—4.5 $1/cm^2$ s. Within the 10-year operation period the integral dose will not exceed 250 rad and will not affect the film performance.

The Van Allen belt protons and electrons exhibit a wide spectrum of energies. A maximum value of the electron flux density is observed at an altitude of 2750 m whereas the proton flux density is increasing as the altitude increases.

At altitudes of 1000—5000 km the cumulative absorbed dose is mainly contributed at the expense of electrons. The dose accumulation velocity and the cumulative dose for circular orbits of similar altitudes are given in Table 1. This Table also presents data on integral fluxes of protons of 40—500 KeV.

*Table 1*

| Orbit altitude, km | 1000 | 2000 | 2750 | 4000 | 5000 |
|---|---|---|---|---|---|
| Dose accumulation; rad/day | $4 \cdot 10^5$ | $4 \cdot 10^5$ | $5 \cdot 10^7$ | $2 \cdot 10^7$ | $5 \cdot 10^6$ |
| including at the expense of electrons | $1 \cdot 10^3$ | $2 \cdot 10^4$ | $8 \cdot 10^5$ | $1 \cdot 10^6$ | $2 \cdot 10^6$ |
| Total cumulative dose, rad | | | | | |
| In 5 years | $7 \cdot 10^8$ | $7 \cdot 10^9$ | $1 \cdot 10^{11}$ | $3.5 \cdot 10^{10}$ | $1 \cdot 10^{10}$ |
| In 10 years | $1.4 \cdot 10^9$ | $1.4 \cdot 10^{10}$ | $2 \cdot 10^{11}$ | $7 \cdot 10^{10}$ | $2 \cdot 10^{10}$ |
| Proton flux density, $1/cm^2 \cdot s$ | $4 \cdot 10^2$ | $5 \cdot 10^4$ | $1 \cdot 10^5$ | $2 \cdot 10^6$ | $1 \cdot 10^7$ |
| Integral flux of protons, $1/cm^2$ | | | | | |
| In 5 years | $7 \cdot 10^{10}$ | $8 \cdot 10^{12}$ | $1.6 \cdot 10^{13}$ | $3.2 \cdot 10^{14}$ | $1.6 \cdot 10^{15}$ |
| In 10 years | $1.3 \cdot 10^{11}$ | $1.6 \cdot 10^{13}$ | $3.2 \cdot 10^{13}$ | $6.4 \cdot 10^{14}$ | $3.2 \cdot 10^{15}$ |

No data are available on how the reflection factor and indicatrix are changing under radiation.

Thus, operational conditions for the reflector film material are peculiar in the combined impact of operational factors (a permanent load at a level of the material yield stress and temperature periodically varying from minus to ≈ 100°C) and space exposure factors (vacuum, radiation, residual atmosphere). These effects are synergistic in nature, i. e. two or more factors acting simultaneously produce different results against their alternate or enhanced effects. Therefore, the most correct results are obtained from the integrated research in materials simultaneously exposed to the UV-radiation and effects of protons and electrons with energies simulating the cumulative dose distribution across the material thickness under stress and thermocycling conditions. No data are available on such experiments.

Of a great practical interest is data on the stability of polymeric materials in space environment. The data has been amassed within the period more than 30 years while using these materials in different space programs. However, by analysing such data, the most often assessment could be made for any one or several factors being the most critical against the combined exposure to all factors. For example, data is available on the environmental exposure of the multilayer thermal insulation package for two months in low orbits. As a consequence, the first 3—4 layers of Mylar film have transformed to grey matter crumbling at the lightest touch. The data are insufficient to conclude on the process kinetics, i. e. in several hours a certain deterioration of the film strength properties was observed and, besides the other data, there was a reason to consider possible the use of this material within the period of up to 2 days. A similar conclusion has been made for the polyimide films.

During the Znamya-2 experiment a TV-camera was viewing the Mylar film surface at a distance of 5—6 m within the period of 4 orbits. The film material was loaded from centrifugal forces, oxygen flux, solar heat, and UV radiation. Within the last orbit, on the TV display, a stress yield effect of the reflector material was observable that was evidenced by an increased curvature of lateral sides of certain film sectors composing the reflector.

The aforesaid factors indicate that an extremely careful approach is required for the selection of film material strength properties governing the structure service life, its controllability in space, surface reflectance quality.

## 5.2. Film Materials and Their Properties

Among films based on polymers are: polyolefine, polyvinylchloride, polystyrene, polyethyleneterephtalate, and others.

Films from 0.005 to 0.1 mm thick are produced mainly by casting a solution or polymer dispersion to polished surfaces, by extruding melts, by wet moulding, and calendering. The US Du Pont Company is producing the Mylar film which is similar to the polyethyleneterephtalate film (PET) produced in Russia and the Kapton film which is similar to the polyimide film (PI) which is also produced in Russia. Nitto Electric and UBE Industries Companies (Japan) have developed polymer-based films (trade marks "Y" and "Upilex") similar to Kapton, however, offering enhanced performance due to a combination of characteristics. Proceeding from physical-mechanical properties and technological effectiveness, as well as considering recommendations from national and foreign experts, those two materials have been selected for developing space-based reflectors. The conclusion has been made that the Mylar film possesses higher mechanical characteristics (elastic strain limit, ultimate strength, tear resistance) within the temperature range up to 200°C against the Kapton film, about the same radiation resistance, higher resistance to UV radiation, and differs greatly in thermal resistance. The Mylar film is recommended for use in space environment within the temperature range up to 200°C. A temperature range recommended for the Kapton film is up to 300°C. By metallizing only a moderate improvement to strength and radiation performance of films could be expected.

Polyimide films of high thermal stability in combination with a high radiation resistance are finding a wide spectrum of applications in different areas including aerospace technology. In implementing the Apollo space program the Mylar aluminium metallized film has been used as a protective coating in fabricating space suits for extravehicular activities, lunar modules of the Apollo-10 and 11 space vehicles, and insulation of the lunar cabin wiring. Principal mechanical characteristics of the Mylar and Kapton films observed under nominal conditions are presented in Table 2.

The strength parameters (ultimate elastic strain $\sigma_s$ and ultimate strength $\sigma_b$) of the Kapton film are given in Table 3.

Thermal radiation properties of metallized films are established for:
- solar radiation absorptivity, $A_s$,
- total normal level of blackness, $\varepsilon_0$.

*Table 2*

| Name of characteristic | Measuring unit | Material | |
|---|---|---|---|
| | | Mylar | Kapton |
| Density | kg/m$^3$ | $1.38 \cdot 10^3$ | $1.36 - 1.42 \cdot 10^3$ |
| Breaking stress:<br>longitudinal<br>lateral | N/m$^2$ | $2.120 \cdot 10^8$<br>$2.78 \cdot 10^8$ | $1.310 \cdot 10^8$<br>$1.320 \cdot 10^8$ |
| Relative elongation ultimate<br>longitudinal<br>lateral | % | 113<br>85 | 17<br>12 |
| Tear resistance (12 m thick)<br>longitudinal<br>lateral | N/m | 5.0<br>6.0 | 1.4<br>1.3 |
| Modulus of elasticity | N/m$^2$ | $3 \cdot 10^9$ | $3.1 \cdot 10^9$ |

*Table 3*

| N | Grade | Thick, $\mu$ | Direction | $\sigma_s$, N/m$^2$ | $\varepsilon$, % | $\sigma_b$, N/m$^2$ | $\varepsilon$, % |
|---|---|---|---|---|---|---|---|
| 1 | Kapton | 8 | longitud.<br>lateral | $25 \cdot 10^7$<br>$2.5 \cdot 10^7$ | 1.8<br>1.8 | $7.11 \cdot 10^7$<br>$7.55 \cdot 10^7$ | 20<br>22 |
| 2 | Kapton | 12 | longitud.<br>lateral | $2.5 \cdot 10^7$<br>$2.5 \cdot 10^7$ | 1.5<br>1.5 | $1.041 \cdot 10^8$<br>$0.917 \cdot 10^8$ | 30<br>29 |
| 3 | Kapton | 20 | longitud.<br>lateral | $2.5 \cdot 10^7$<br>$2.5 \cdot 10^7$ | 1.3<br>13 | $1.06 \cdot 10^8$<br>$1.023 \cdot 10^8$ | 50<br>48 |

| Parameter | Surface state | Mylar | Kapton |
|---|---|---|---|
| $A_s$ | Metallized | 0.15 | $0.13 - 0.15$ |
| $\varepsilon_0$ | | $0.05 - 0.06$ | $0.03 - 0.04$ |
| $A_s$ | Non-metallized | 0.07 | $0.28 - 0.31$ |
| $\varepsilon_s$ | | 0.2 | $0.49 - 0.53$ |

## 5.3. Experimental Determination of Metallized Film Reflection Factor in Two-Dimensional Stress State

In using the space-based reflector the main objective is to provide a maximum level of illumination of the region being served. The illumination level depends on the reflector area, orbit altitude, Sun's position, angle, place, and varies in a direct proportion to the reflection factor.

In performing calculations for the space-based reflector by the mirror reflection factor is meant the ratio between the reflected sunlight within the angle of $31'59'' \pm 32''$ (0.00931 rad) and the total incident flux. The angle $31'59'' \pm 32''$ (0.00931 rad) corresponds to the sunlight angular divergence.

In a general case a mean value, $(\chi)$, of the reflection factor applied to the entire area of the reflector we may write as:

$$\chi = \frac{1}{S} \int \chi_n \cdot \chi_{ph} \cdot \chi_m \cdot \chi_{sl} \cdot dS$$

where $S$ — area of space-based reflector;
$\chi_n$ — mirror reflection factor of reflecting coating;
$\chi_{ph}$ — mirror reflection factor of space-based reflector dynamic macroform;
$\chi_m$ — mirror reflection factor of material;
$\chi_{sl}$ — mirror reflection factor of space-based reflector static macroform.

$$\chi_n = 1 - \varepsilon - q$$

where $\varepsilon$ — coating blackness level;
$q$ — transmission factor.

For aluminium coatings of polymeric films value $\chi_n$ equals 0.90...0.95 and depends on the coating thickness and continuity, i. e. a value of q.

For Al $\varepsilon = 0.05$.

$\chi_{ph}$ — is determined from the surface deformation of the space reflector in the process of its attitude control. This parameter is not considered in the given case.

$\chi_{sl}$ — is determined from tolerances for cutting out patterns and can be taken equal to one at angles below 0.16′ between separate areas of the space reflector and its base plane in a stress state.

Proceeding from the operational conditions and dedicated use of film reflectors built by centrifugal forces, technical requirements have been established for film materials. One of the requirements was to ensure the reflection factor within 0.6—0.7.

A special experimental setup has been designed and fabricated to generate a two-dimensional stress state of film materials with respect to elastic and plastic strains.

The reflection factor of polyimide films 12 $\mu$ thick metallized has been experimentally studied. Studies have been conducted on how the reflection factor depends on the film tension. Experiments have been also conducted to explore the presence of folds and seams on the film material at three angles of sunlight incidence.

The obtained results refer to the experimental determination of $\chi_m$ or more specifically, to products $\chi_n \cdot \chi_m$. But, as far as $\chi_n$ is close to 1, the results could be applied to $\chi_m$. The measured $\chi_m$ is characterising the material reflection factor in a small scale (a diameter of the area under research was $d_2 = 30$ mm).

### 5.3.1. Experimental Setup Description

Since the metallized polymeric films are intended to be used within the real sunlight flux, a setup for measuring the mirror reflection factor should meet the requirements for measuring conditions and actual operation environment.

An optical scheme of the experimental setup is shown in Fig. 5.1. The light flux generated by sunlight simulator (1) is directed at angle $\alpha_n$ to the film sample installed in special device (2) providing a controllable, uniform tension. The light flux reflected from the sample is recorded by measuring photometer (3) placed at the same angle $\alpha_n$ relative to the normal to the sample surface.

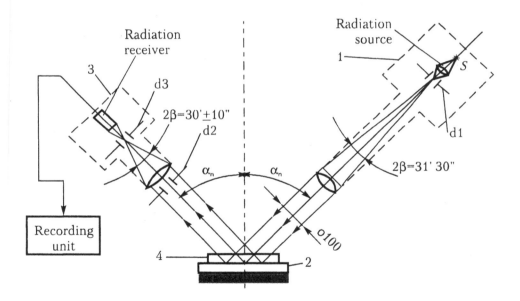

Fig. 5.1. Setup optical diagram. 1 — mirror reflection factor optical measuring scheme; 2 — radiation receiver; 3 — radiation source; 4 — recording unit

The Sun simulator uses a projection lamp is used in the capacity of light generator $S$. The light flux generated by the lamp, having passed through a lens condenser and membrane $d_1$, is focused by the two-lens objective with a focal length of $f = 1000$ mm. A diameter of the outcoming radiation beam is 100 mm, an angular divergence — $2\beta$ = 31'30" (determined from the membrane diameter, $d_1$) that corresponds to a mean angular size of the real Sun.

To limit the measured surface area of film samples, membrane $d_2$ 30 mm in diameter is placed in front of measuring photometer lens (3) to restrict the measured light flux.

The light flux restricted by membrane $d_2$ is focused by the photometer lens within the plane of membrane $d_3$ which establishes the angular field of view $2\beta$ of the photometer at a level of 30'—10".

Behind membrane $d_3$ a photomultiplier photodetector is installed to send electric signals to the recorder.

To direct the light flux accurately to measuring membrane $d_3$, the photometer is provided with a visual observation channel which eye glass serves to control the accuracy of making coincident the "Sun" disk with measuring membrane $d_3$. The Sun simulator and the measuring photometer are installed in special coordinate-rotary devices allowing to set angles $\alpha_n$ at an accuracy of $\pm 1'$ (ang.).

To provide the stability control and alignment of the light flux radiator (the Sun simulator) prior to each of the measuring cycles a calibrating optical flat mirror has been used (4). The mirror reflection factor was measured on the same setup for angles $\alpha_n$ = 30, 45, 60. The reflection factor of the calibrating mirror was not depending on angles $\alpha_n$ and was equal to 0.906.

An experimental setup has been designed to study the reflection performance of coatings on film materials with their tension being monitored and controlled.

The setup provides the uniform two-dimensional stress state controllable in the area of elastic and plastic strains of polymer-based film materials (Kapton, Mylar, etc.).

The block-diagram is shown in Fig. 5.2.

Operation flow: film under study (1) is fastened by means of mounting flanges (3), (4) providing the initial smoothing of the film sample using the principle of stretching on a "tambour-frame" followed by clamping the film over the working planes. By rotating spring-loaded flywheel (7) pressing flange (2) is brought to the film. A force generated by calibrating compression member (8) is recorded in the direction of motion indicator (6). A height of lift of the pressing flange is recorded in the direction of motion of the pressing flange indicator stud (5).

The studies have been conducted on the polyimide film sample having one-side metal coating (the ПM-IЭУ-OA TУ17, Latvia, 0306-87 grade similar to Kapton) 12 μ thick.


### 5.3.2. Optical Measurements Techniques

The polyimide film reflectance versus its tension has been studied using the aforesaid experimental setup. The reflection factor was measured at three incidence angles of light radiation $\alpha_n$ = 30°, 45°, 60°. Angle $\alpha_n$ was counted off from the normal to the film sample surface (see Fig. 5.1).

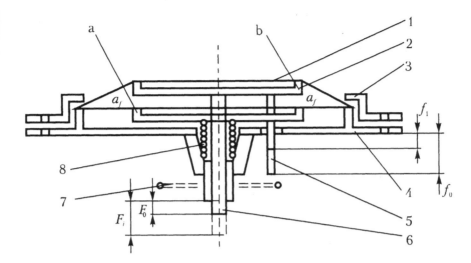

Fig. 5.2. Two-dimensional film tension device outline. Principal scheme of installation. a) initial position of pressing flange 9; b) pressing flange in service. 1 — tested film; 2 — pressing flange; 3 — upper mounting flange; 4 — lower mounting flange; 5 — pressing flange motion indicator; 6 — compression member stroke indicator; 7 — spring-loaded wheel; 8 — compression member

The experimental film sample was fastened within the setup in the direction of its tension (Fig. 5.2). Then, by rotating the spring-loaded flywheel, the film was subjected to the initial tension with a view to avoid errors in calculating the film tension. Such errors could result from inaccuracies crept during the initial fastening of the film sample within the experimental setup, improper preliminary smoothing of the film sample, large wrinkles, and initial slack. The initial tension was ascertained by the commencement of pressing motions of the compression member responsible for the film tension.

To control the light flux reflector (the Sun simulator) stability and to provide the alignment, the calibrating, optical, flat mirror (4) was used prior to each of the measuring cycles. The reflection factor of the mirror was measured on the same experimental setup for angles $\alpha_n$ = 30°, 45°, 60° and appeared to be $C_m$ = 0.906.

Thus, the mirror reflection factor is determined from the ratio between the radiant flux reflected from the experimental sample and the radiant flux incident to the sample. The reflection factor is measured through determining the relation between the radiation receiver photocurrents varying in the direct proportion to the reflected and incident radiant fluxes:

$$C_m = \frac{I''}{I'}$$

where $I''$ — photocurrent caused by the reflected radiant flux;

$I'$ — photocurrent caused by the incident radiant current.

Through rotating the spring-loaded flywheel the film was subjected to further tension with the established pitch. Readings were taken for each pitch from photometer $\Gamma'$, motion indicators of compression member $F$, and lift indicators of spring-loaded flange $f$. The reflection factor has been calculated under the above techniques thereafter.

### 5.3.3. Results of Measurements

1. Maximum values of the reflection factor for the commercial, one-side metallized, pure polyimide film 12 $\mu$ thick are within a range of 0.58—0.66 (Fig. 5.3).

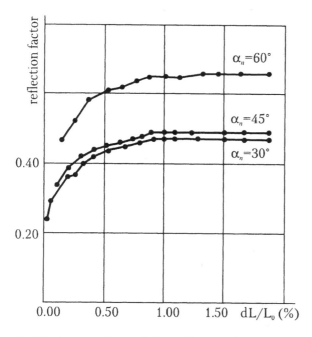

Fig. 5.3. Mirror reflection factor measurement results

2. Minimum values of the reflection factor for the commercial, one-side metallized, pure polyimide film 12 $\mu$ thick are within a range of 0.465—0.471.

3. Maximum values of the reflection factor for the randomly crumpled, commercial, one-side metallized, polyimide film 12$\mu$ thick are within a range of 0.22—0.23.

4. Minimum values of the reflection factor for the randomly crumpled, commercial, one-side metallized, polyimide film 12 $\mu$ thick are within a range of 0.12—0.13.

5. The reflection factor of polyimide film increases proportionally to the light flux incidence angle measured from the normal to the film surface.

6. The reflection factor of polyimide film increases proportionally to its tension (tensile elongation) approaching to the maximum value (Fig. 5.3).

7. With the film tightened in the plastic strain area the reflection factor is gradually decreased.

8. Seams contribute greatly to the reflection factor decrease.

9. Wrinkles have only a slight effect on the reflection factor.

10. In scanning the film sheet along the seam in the direction of the light flux a lower reflection factor was observed as compared to scanning along the seam normal to the light flux.

11. With a wrinkle 10 mm wide its impact to the reflection factor is propagated approximately over 100 mm in the direction of the reflection factor reduction.

12. The reflection factor of the sample containing the cross seam was slightly lower in a point 50 mm away comparing to that of the solid sample.

13. In loading the film a time period is estimated for setting the tension within which the reflection factor is varying in a range of $\pm 1\%$.

14. The elastic strain of the experimental film sample falls within a range of $\approx 0.8\%$. The value increases at repeated loading.

15. Seams and wrinkles are contributing to the elastic strain of the film sample.

Thus, it is seen from the experiments that:

• two-dimensional uniform tension of film contributes largely to its reflectance,

• by increasing the light flux incidence angle the film reflectance is improved,

• two-dimensional uniform tension of film allows to smooth out wrinkles and to minimise their impact on the film surface reflectance.

## 5.4. Experimental and Theoretical Research in Creep and Long-Term Strength of Polyimide Films

### 5.4.1. Durability Assessment Technique

Experimental research has been conducted in creep, long-term strength, and fatigue of polyimide films. The structural-phenomenological model has been developed to assess and predict the durability at the varying cyclic temperature and constant loading. The procedure for express durability tests under the aforesaid conditions has been proposed. The work objective was to substantiate the selection of strength properties of film materials intended for space-based reflectors of a durable service life peculiar in cyclic changes of temperature due to the variable in-orbit orientation of the reflector relative to the Sun and a constant load. To assess creep in the case of thin polymeric films at the cyclic changes of temperature the following equations are proposed:

$$\varepsilon_c = (1/cm)^{1/m} \left[ T_m(1+a\tau^u) - (T_u - T_l)1/m + (T_u - T_l) \right] / E ,$$

where  $T_u$  — upper temperature of the cycle;

$T_l$  — lower temperature of the cycle;

$T_m$  — mean temperature of the cycle;

$c, m, a, u$ — constants of the material determined form the instant temperature strain
diagram and the creep curve.

Relying on the experiments it had been concluded that the creep strain would be
constant on destruction and would range between 70—80% from the breaking strain
at the short time loading.

The time to destruction, i. e. the durability, is calculated from the following equation:

$$\tau_d = B_0 / T_m (A - (T_u - T_l)/E)^m,$$

*where:*

$$B_0 = a/c\,m\,; \quad A = |\varepsilon_p|$$

The creep velocity is given by:

$$\varepsilon_c = a\,u\,(T_m/E) \cdot 1/\tau^{1-n}.$$

### 5.4.2. Experimental Setup and Test Methods

Uniaxial tension tests of film samples have been conducted on the modified test
setup. Considering that in the real operational environment the film materials work
under the two-dimensional stress, a special fixture has been designed, fabricated, and
verified to fasten circular samples of thin films and to test them under the inside gas
pressure (argon) within a range from 0.1 to 10 atm. (Fig. 5.4). The tested sample
strain was measured with a clock-type indicator placed in the middle of the sample.

The test temperature was maintained by means of the rheostat-type heaters. The
transition to the mode of up to 100°C had been performed in 20 minutes and the
specified temperature was maintained at an error of ±5°C within a durable period of
time. The temperature was controlled with two thermocouples of chromel-alumel in
places 10 mm away from fasteners.

To conduct cyclic tests, an automatic setup had been employed that allowed to
maintain the cyclic temperature.

The following tests have been conducted under the uniaxial tension and two-dimensional stress from the inside pressure:

• short time tests to determine the breaking stress and tensile elongation,

• long tests to plot the long-term strength diagrams and creep curves at constant
stationary loads,

• cyclic tests (non-stationary) under stepped loads to assess the long-term
strength and creep.

Table 4 contains parameters of short testing cycles.

Fig. 5.4 Setup for film creep investigation

*Table 4*

| Mode | $\tau_1$ (heating) | $\tau_2$ (exposure) | $\tau_3$ (cooling) | $\tau_4$ (heating) |
|------|--------------------|---------------------|--------------------|--------------------|
| I | 20 min | 20 min | 2−20 min | 20 min |
|   | 2 hours | 2 hours | 4 hours | 16 hours |
|   | 8 hours | 8 hours | 4 hours | 4 hours |
| II | 20 min | 0 | 2−20 min | 20 min |
|   | 2 hours | 0 | 4 hours | 18 hours |
|   | 8 hours | 0 | 4 hours | 12 hours |

### 5.4.3. Test Results

Results of the creep tests conducted in mode I are shown in Fig. 5.5 at $T_{max}$=75°C over 500 hours. The data was used to make predictions for the periods of 5000 and 10000 hours.

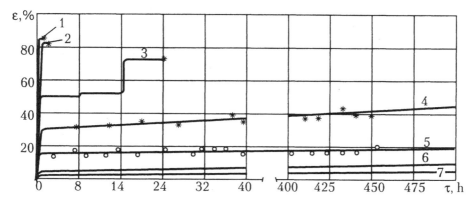

Fig. 5.5. Creep of samples at $T_{max}$ = 75°C. $1 - \sigma_1 = 0.8\,\sigma_p$; $2 - \sigma_2 = 0.5\,\sigma_p$; $3 - \sigma_3 = 0.35\,\sigma_p$; $4 - \sigma_4 = 0.3\,\sigma_p$; $5 - \sigma_5 = 0.25\,\sigma_p$; $6 - \sigma_6 = 0.2\,\sigma_p$; $7 - \sigma_7 = 0.11\sigma_p$. Tests No $1 - 3$ completed in destruction; tests No $4 - 7$ no destruction of samples

Short tests for durability (Fig. 5.6) have been conducted in the steady loading mode under 20°C, 75°C, 100°C (curves 1, 2, 3) and also under cyclic temperature changes at $T_{max}$ =75°C, 100°C (curves 4, 5).

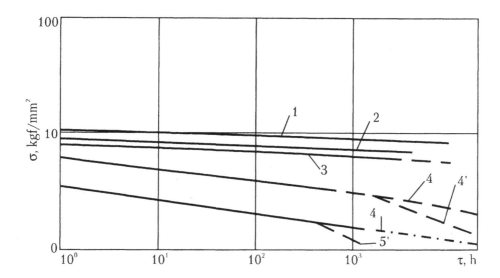

Fig. 5.6. Durability of samples in stationary and cyclic temperature tests

The predictions ware made in three steps.

During step I the breaking strain was evaluated. The creep strain steadiness at breaking was taken as the performance criterion, i. e. $\varepsilon_p$= const at the maximum estimated cycle temperature. The experimental data analysis showed that:

• the breaking strain during long tests at $T_{\text{test}}$= 20°C under the steady, two-step, and three-step loading was approximately of the same value, $\varepsilon_p$ = 40—55%,

• the breaking strain under the uniform heating from 20°C to 100°C and different loads acting on the samples was actually constant, $\varepsilon_p$ = 75%,

• during short tests $T_{\text{test}}$= 75°C, $\varepsilon$ = 80%, during long cyclic temperature tests in mode I $\varepsilon_p$= 70%.

For the durability margin the lowest value of $\varepsilon_p$ obtained in different loading modes at the specified maximum temperature must be considered.

During step II the short, non-destructive, cyclic tests were conducted (curves 4, 5, Fig. 5.6) and the durability curve was plotted (for 300—400 h). The tests were terminated after the constant creep velocity had been set.

At step II the long strength limit was determined for the established durability, e.g. for 5000 or 10000 hours.

The prediction results are given in Fig 5.6 (curves 4', 5').

The given results are lower comparing to those obtained from ratios of long-term strength used in forecasting metallic materials. It should be emphasised that life tests of metallic materials are associated with a large scope of long-term tests conducted in parallel on a large number of samples. Such tests have to be continued with a view to collect statistics, to increase durability using solid samples, and to repeat the entire work scope using welded and bonded seams.

## 5.5. Radiation Resistance of Film Materials

It is customary to evaluate the material performance from changes of the material properties under exposure to fast electrons. Experimental data on changes of the mechanical ($\sigma_p$) and optical ($R_s$) properties of the polyimide film exposed to electrons of 5·10 MeV are given in Fig. 5.7a versus the radiation dose. Under the real operational conditions the radiation absorbed dose is mainly contributed by electrons with energies of dozens and hundreds of KeV and the lower velocity of dose accumulation. The metallized film properties could change in a radically different manner, namely, deteriorate against the results shown in Fig. 5.7a.

Data is available on how protons ($E$ = 500 KeV) affect the metallized polyimide film mechanical properties and how the UV radiation affect the mechanical properties (up to 200 ESD) and the solar radiation integral reflection factor (up to 600 ESD). The data is given in Figs. 5.7b and 5.7c.

The Mylar film is tolerant to electron radiation of high (over 1 MeV) energies and, actually, remains unchanged in performance while intensively absorbing a short-wave portion of radiation. Data is presented on the laboratory experiments to

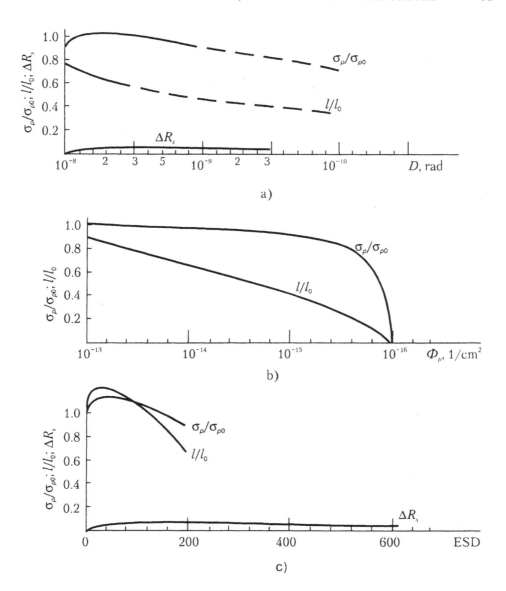

Fig. 5.7. a) Change of metallized Kevlar film exposed to electrons; b) change of metallized Kevlar film exposed to protons; c) change of metallized Kevlar film exposed to the solar electromagnetic radiation

ascertain the film destruction caused by the UV radiation and accompanied by the formation of numerous gaseous products including toluene and benzaldehyde.

Fig. 5.8 presents the summarised data on the deterioration of strength and optical properties of the Mylar and Kapton films exposed to the UV radiation.

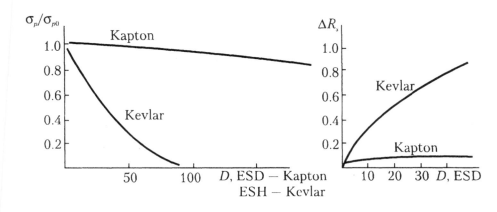

Fig. 5.8

For the Kapton film the radiation dose is given in Equivalent Solar Days (ESD) and for the Mylar film — in Equivalent Solar Hours (ESH). The strength properties are given by ratio $\sigma_p / \sigma_{p_0}$ between the breaking stress under the UV radiation and the initial breaking stress, $R_s$ (reflectance). The Kapton film demonstrates a significantly higher resistance to the UV radiation. Only a moderate improvement of the aforesaid properties would be achieved through the aluminium coating.

## 5.6. Designs of Film Structures

Principal requirements to be met in designing film structures:
  • a controlled deployment,
  • an in-process packaging,
  • a minimum number of seams per unit of surface area,
  • the best use of the original film material (minimum of wastes because of a high cost of the film),
  • the use of templates and limited production areas for cutting and combining the film sheets (bonding or welding, automatic or manual),
  • an easy evacuation of air out of any stowage volume, no gas entrapped in a stowed structure.

Historically, the first film structure employing the bellow-type folding pattern (Fig. 3.1) was meeting, to the greatest extent, the requirements listed above in par. 2—6, as compared with all subsequent structures has been analysed at the Dolgoprudni Automatics Design Office (DADO). However, as it emerged later, this structure failed to meet the first requirement. The follow-on developmental effort has been focused on the folding pattern as per ref. [2] (Fig. 3.2). Technical documentation has been developed for the design and production technology of a film sheet of

100 m in diameter; several Mylar sheets of 5 m and 20 m in diameter have been fabricated. In verifying the production flow the pattern as per ref. [2] appeared to be low-effective and arduous. The most intricate was to make the last longitudinal seam due to fed on tolerances.

During the third developmental stage the split sheets have been evaluated. This design was much simplified and efficient comparing to the solid canvas. Three sets of spools containing split sheet of $D$ = 25 m and two sets of spools containing a split sheet of $D$ = 20 m have been fabricated for the Znamya-2 experiment. For a separate sector the "down-up" folding pattern was derived from the structure as per ref. [2] with the only difference that the fold run in the radial instead of tangential direction and each of the sectors was reeled up on its spool. This structure has been successfully deployed during the space-based Znamya-2 experiment. However, the DADO experts recognised that there was no point in using the "down-up" folding pattern of the structure as per ref. [2] in future split structures for reason of the low efficiency offered by split sheets. As a consequence, a split reflector is now under development which will be fabricated of sectors combined of different strips (Fig. 5.9). This reflector will not require an extremely arduous, manual process of smoothing out angles of the "down-up" folding pattern (Figs. 3.2, 5.10). Presently, to design the solid reflector the DADO structure proposed as early as in 1989 for a solar sail is under consideration. (Fig. 3.4). The structure is combined of the pre-fabricated sheets of tangentially arranged film strips. The strips are pre-folded into the bellow-type pattern parallel to seams in the process of their fabrication and the bellow-type pattern elements are coupled in the process of the final assembly and reeling up on the central cylinder (Fig. 3.5 3.7). The middle of each strip is pulled to the central cylinder by means of a tether fed through a hole in the centre of each strip (Fig. 5.11). The tether is employed for the strip controlled deployment thereafter.

A drawback of this design is that, during deployment, the sector structure would inevitably slide over the tether with an unknown friction thereby causing the possibility of sticking due to the tether micro-roughness, or cutting the eyelets of the strips. The option is under consideration where a loop of two tethers is thrown over the stowed structure. The external loop assures the folded structure from sliding over the tether during deployment, however, due to the asymmetric configuration of the structure caused by tolerances, it appears probable that the structure would loose its stability, the probability which, in turn, would become less dangerous with a larger number of sectors ($n$ = 8, 12, 16).

Presently, to decide the final folding pattern, the ground simulation of the structure deployment is performed to verify the process. Of the four possible patterns considered for folding the "tails" ("serpentine", "in bag", on separate spools, coiled on a central cylinder) the coil pattern could be favoured after detailed studies of both the film sheet structure and its deployment mechanism. This pattern is the maximum simple from the engineering viewpoint and would not require the second layer of tethers for the controlled deployment (that, actually, made twice as complicated the deployment mechanism structure due to a necessity to fully duplicate the tethers, spools, motors) and would not pose the problem of disposing the container and tethers of the first layer upon the structure deployment. As far as the dynamics of unreeling the "tails" from the coil pattern is not completely controllable, all the com-

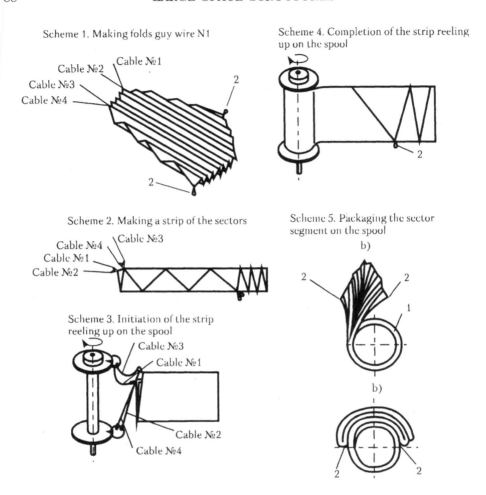

Scheme 1. Making folds guy wire N1

Cable №2
Cable №1
Cable №3
Cable №4
2
2

Scheme 4. Completion of the strip reeling up on the spool
2

Scheme 2. Making a strip of the sectors

Cable №4
Cable №3
Cable №1
Cable №2

Scheme 3. Initiation of the strip reeling up on the spool
Cable №3
Cable №1
Cable №2
Cable №4

Scheme 5. Packaging the sector segment on the spool
b)
2
2
1

b)
2
2

Fig. 5.9. Packaging of the split sheet sectors

plicity of this process is shifted to the tether flywheel, i. e. in the task involving three rotating bodies (reflector, vehicle body, and flywheel) the deployment will be controlled at the expense of the rotation drive performance and the tether flywheel deployment velocity. The controlled deployment of the "tails" through the peripheral belting or catching each "tail" with the tether also required the duplication of the deployment mechanism structure and involved the unknown process of sliding of the film rolls relative to each other while being deployed. This unexplored sliding process should be inevitably attended with scores on seams, deterioration of the reflecting surface performance, possible displacement of the entire package in any direction resulting in a disbalance.

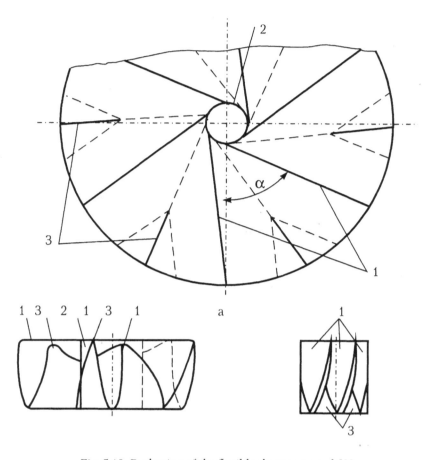

Fig. 5.10. Packaging of the flexible sheet as per ref. [2]

The structure of sheets expanded by centrifugal forces incorporates tether elements providing the guying-type fastening to the central cylinder (Fig. 1.2), both in the split and solid designs. The split design also contains tether elements for joining sectors to one another.

In certain designs a supporting structure is used to keep the film material rigid. Among such designs is the "heliorotor" design [1] where centrifugal forces keep the structure expanded (blades 8 km long and 4 km wide are made of the Kapton film stretched over a metal spring frame). The "heliorotor" structure was a prototype of a solar sailing vehicle where a rigid frame was replaced by linking the sectors-blades around the periphery thus preventing the blades from torsional oscillations being a substantial drawback of the "geliorotor". The solar square-sail 4000 $m^2$ in area jointly designed by France and Spain is made of the Kapton film 8 μm thick (Fig. 2.1). The film sheet is folded into a square configuration in two perpendicular directions following the bellow-type pattern. While making the final package, the stowed structure is evacuated and an extremely high packaging coefficient, close

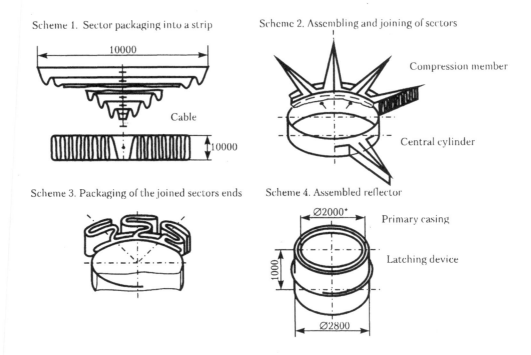

Fig. 5.11. Film sheet fabrication and packaging technology

to 1, is achieved (a ratio between the stowed volume and the film material volume). This square sail is being fabricated at the DuPont facilities using the DuPont equipment.

To sum up this Chapter, it should be remarked that the film structure could be selected only by comprehensively analysing the design, engineering, dynamic, and operational aspects related both to the film material and its deployment mechanism.


## 5.7. Selection of Strength Characteristics. Surface Accuracy

It is worth noting that the DADO technology affords a low production accuracy (4 mm tolerance for 1 m). The experience is known in manufacturing film structures for telescope surfaces at an accuracy exceeding by one order of magnitude that of the structures of standard sizes up to 15 m. The Contravers company (Zurich) is employing the pre-tension technology and automatic laying out to manufacture the film parabolic mirror of 6 m diameter at an accuracy of 0.8 mm, whereas the JPL company (US) is making the film parabolic mirror of 15 m diameter at an accuracy of 1.6—1.8 mm.

Studies on the metallized film reflection performance had been performed which showed a reflection coefficient running to a maximum at the film material strain close to a yield strength, i. e. at a strain of the order of 1.5—1.8%, that is nearly four times as much as errors associated with the production accuracy. There, irregularities related to the production accuracy, in fact, will not affect the reflection performance.

Consideration is also given to the fact that the structure controllability requirements and thereby introduced macro-irregularities (for example, a circularly running wave caused by gyroscopic forces) necessitate a high tension of film material in the reflector structure built by centrifugal forces. It thus appears that the limit exists beyond which there is no need to make the film tension lower and, as a consequence, to improve the film reflector production accuracy. The aforesaid pertains equally to the framed structures where a necessity of film tension to ensure its reflection performance requires rather high mass-dimensional properties of the supporting structure to make it sufficiently rigid. However, dynamic impacts associated with the object re-orientation in space and causing the surface macro-irregularities are less in the framed structures as compared to the structures built by centrifugal forces. It therefore suggests that with the improved technology of applying coatings and measures affording a high quality underlay, a high reflection factor would be obtained at lower strain values of the underlay material. Consequently, this would allow to reduce the supporting structure rigidity and its mass-dimensional properties and lead to a necessity of improving the reflector surface production technology.

## 5.8. Reflector of Rotating, Uniformly Stressed Film Sheet

The problem of designing the large space-based reflector built by centrifugal forces and dedicated for different application specific tasks is dealing with the desired maximum reflectance. The results of the ground experiments for determining the reflection factor of metallized films suggested that, when subjected to a uniform, two-dimensional stress, the reflection factor would gradually increase with the tension increased and would run to the maximum value (0.75—0.8) at the end of the elastic strain zone. With the tension further increased the reflection factor would start to decrease. Thus, the reflector should be operated at a rather large, two-dimensional, uniform tension close to the yield strength limit considering the long-term and creep.

The assessments indicated that with the non-profiled film sheet only 30% of the reflector area would provide the optimal reflection capability owing to an essential difference between radial and tangential stresses in the rotating film sheet of a uniform thickness. This is unacceptable from the viewpoint of the reflector assumed use in the application specific systems. The reflector structure profiled along its radius offers the solution to this problem. Based on the solution of the task of the uniformly stressed disc where radial and tangential stresses are uniform over the entire area of the rotating reflector structure, the required angular velocity of rotation and the law of the reflector thickness profile variation could be established. The law of the reflector thickness profile could be obtained via the graduated selection of film fragments from the available variety of films as the reflector is being fabricated. There, two methods should be combined:

• a weak profiling when different fragments are selected from one lot of film tolerated to 20% in thickness;

• a strong profiling when lots of film essentially different in thickness are used.

The initial combined equations for the composite-stressed state of the rotating disc of a thickness variable along the radius is given as:

$$\frac{r}{h}\frac{d}{dr}(h\sigma_r) + \sigma_r - \sigma_\varphi + \rho\omega^2 r^2 = 0 \tag{5.1}$$

$$\frac{d}{dr}\left[\frac{r}{E}(\sigma_\varphi - \mu\sigma_r) + r\alpha_t\right] = \frac{1}{E}(\sigma_\varphi - \mu\sigma_r) + \alpha_t \tag{5.2}$$

where $h$ — disc thickness;

$\sigma_r, \sigma_\varphi$ — radial and tangential stresses;

$\rho$ — film material density;

$\mu$ — the Poisson's ration;

$t$ — temperature;

$\alpha$ — thermal expansion coefficient;

$E$ — modulus of elasticity.

To obtain maximum reflection capabilities the radial and tangential stresses should be equal and constant along the disc radius, viz

$$\sigma_r = \sigma_\varphi = \sigma = \text{const} \tag{5.3}$$

By substituting (5.3) in (5.1) we obtain:

$$\frac{dh}{h} = -\frac{\rho\omega^2 r}{\sigma}dr \tag{5.4}$$

By integrating both parts of the equation we obtain:

$$\ln h = -\frac{\rho\omega^2 r^2}{2\sigma} + C \tag{5.5}$$

Integration constant $C$ is derived from the boundary conditions, viz

$$r = R_k, h = h_{rk}, C = \ln h + \frac{\rho\omega^2 R_k^2}{2\sigma} \tag{5.6}$$

And, hence, we have:

$$h = h_{R_k} e^{\frac{\rho\omega^2(R_k^2 - r^2)}{2\sigma}} \tag{5.7}$$

This equation of quadratic exponential function is well known in the area of wheel-assisted energy accumulators. The equation demonstrates the laws describing the thickness variation along the radius of the uniformly stressed, rotating disc. For the disc centre at $r = 0$ we have:

$$\frac{h_{r=0}}{h_{R_i}} = e^{\frac{\rho \omega^2 R_i^2}{2\sigma}} \qquad (5.8)$$

From equation (5.8) the lower boundary for the angular velocity $\omega$ can not be obtained. It follows from equation (5.8) at $h_{r=0}/h_R \approx 1$ that is for a disc of a uniform thickness, $\omega$ tends to approach zero that has no physical sense. On the other hand, for a solid disc of a uniform thickness equations (5.1, 5.2) have the known solution for the disc centre at $r = 0$, viz

$$\sigma_r = \sigma_\varphi = (3 + \mu) \rho \omega^2 R_k^2 / 8 \qquad (5.9)$$

From the latter expression the lower boundary for the rotation angular velocity $\omega$ could be determined.

In practice, for the numerical solution of equations (5.1, 5.2) the disc is split into a large number of concentric zones within which the disc thickness is constant and the solution is found by the method of successive approximations when the radial and tangential stresses are actually constant with an appropriate $\omega$ and the disc thickness profile gradually varying along the radius.

It should be noted that the aforesaid is referred to the reflector structures being non-split along the radius.

In the split reflector structure consisting of separate sectors no tangential stresses would actually exist, and to provide such stresses certain design measures could be undertaken, for example, making the sectors profiled and introducing the stiffening members in the sector structure, however, this would result in a loss of advantages offered by structureless systems.

Let us consider the sector structure outline shown in Fig. 5.12. The sector of maximum profiling height $f_1$ along radius $R$ and $f_2$ around periphery $L$ is rotating about axis 0 within the drawing plane. In point A combining two tops sectors mass $m$ is placed to expand two adjacent sectors. If constant, distributed load $q$ acts on the outline, to obtain force $F_1$, mass $m$ should be applied to point A to equalise both constituents of forces acting on the outline which are calculated from the known expressions of structural mechanics for a case of small deflection $f$, viz

$$F_1 = m \omega^2 R = \frac{2 q R^2}{8 f_1} + \frac{2 q L}{2} \qquad (5.10)$$

As an example, let us make the numerical estimation of the Znamya-2 experiment parameters, that is $R = 10$ m; $\omega = 2$ rad/s, film thickness $\delta = 10^{-5}$ m; $\sigma = 10^7$ N/m$^2$, that is $q = \delta \sigma = 10^2$ N/m. Assume $f = 0.1R = 1$m and $L = R$, then $m = 88$ kg and with six sectors — 528 kg. Whereas the entire sheet mass is about 4 kg, it would be reduced to 3.5 kg by increasing the rotation velocity to 10 rad/s. With six sectors the mass

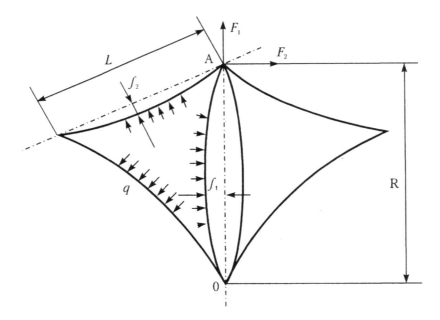

Fig. 5.12. Sector structure

would decrease to 21 kg. The radial stress is preset by profiling the upper part of a sector via the selection of required value $f_2$ that will also reduce the reflector effective area. The sector could be also radially stressed by distributing masses around the periphery and by profiling the film thickness along the radius.

A film sector should be profiled taking account of a strong strain associated with a force impact and an increased temperature of the reflector material. A force creating distributed load could be applied to the sector structure only by a tether having a sufficiently greater modulus of elasticity. There, the motion of the film being expanded relative to the unstretchable tether would be complicated along a large length of profiling.

The more rigorous calculations have been made by solving the combined Equations of the stressed state in partial derivatives by the finite element method. The results appeared more accurate but fundamentally similar to the aforementioned estimations.

As a whole, the feasibility analysis of film reflectors pointed to severe difficulties which would be suffered in building the surface of the two-dimensional, uniform stress state in the case of profiling separate sectors and the unreasonable use of split reflectors.

Solid circular sheet of the film material profiled in thickness along the radius could offer high reflection capabilities and essential advantages against split structures.

# CHAPTER VI

# GROUND EXPERIMENTAL SIMULATION

## 6.1. Experimental Research of Tethered Systems

### 6.1.1. Purpose and Objectives

The purpose of experimental research is to make a physical model of rotating "floppy" systems which could be deployed in air environment within the gravitational field.

The objectives of these studies included:
1) visual observation of the systems,
2) elucidation of systems capable to keep the required, stable configuration.
Fig. 6.1 depicts the schemes containing the experimental tethers.

### 6.1.2. Tethered System Deployment Mechanism Design

In implementing the objectives of the experimental studies the Proton experimental models have been developed and manufactured. The models are designed so that their functional capabilities meet the requirements for certain stages of the studies.

The Proton-4 model (Fig. 6.2) has been developed to study how a tether is moving in the process if its building into a loop of a specified configuration, as well as to explore the loop motion stability while changing its rotation plane.

The model structure is shown in Fig. 6.3. Synchronous electric motor 2 is mounted on cover 3 of housing 1. Cover 3 is simultaneously serving as a casing of re-

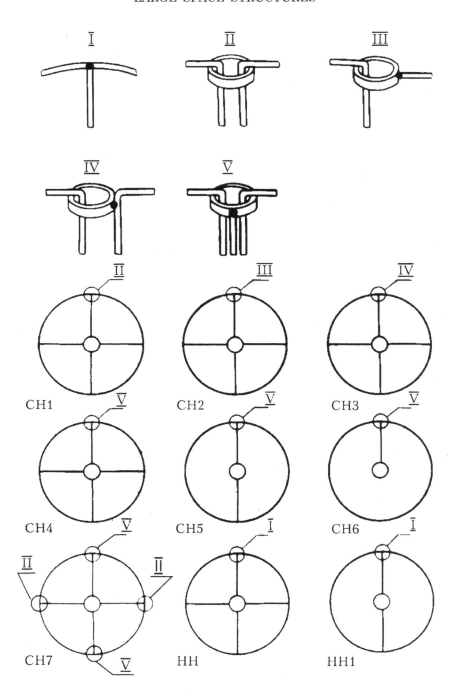

Fig. 6.1. Tether system designs

Fig. 6.2. Proton-4 experimental model

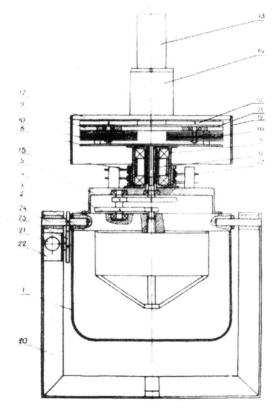

Fig. 6.3. General view of the Proton-4 model

ducer 4. Sleeve 7 put on the reducer output shaft is rotating on rolling bearings fit on stationary sleeve 6 mounted on cover 5 of the reducer. Flange 8 is fastened to the sleeve. Between flanges 8 and 9 joined together through studs 10 the three-layer spools 16 are mounted on the output shafts of one-step reducer 15.

The tether is released by means of electric motor 13 through planetary reducer 14 and reducer 15. The electric motor and the planetary reducer are installed on flange 12 which is fastened to flange 9 through struts 11. Reducer 15 is mounted between flanges 9 and 12. Electric motor 13 is powered through collecting rings 18.

Electric motor 21 is changing the orientation of the tether loop rotation plane via the closed, worm-cylinder reducer 22 and open cylinder reducer 23. A flywheel of reducer 23 is fastened on axis 24 installed on cover 3 of housing 1. Axes 24 are resting on studs of mount 20.

The Proton-4 electric motor is controlled from a remote panel.

The Proton-5 model (Fig. 6.4) has been developed to study the dynamics of the simultaneously counter-rotating tether loops, the release of radial and circumferential tethers being independently controlled. Besides, in designing the model, the possibility was considered to perform the simplest experiments on membranes and coatings.

Fig. 6.4. Proton experimental model design

The model structure is shown in Fig. 6.5. Two similar devices are installed on shaft 1. The rotation is provided through bearings 2 fit on stationary sleeve 3. Flanges 4 and 5 fastened to bearing bodies 6 accommodate primary motors 7 setting the device in rotation via gear transmission 8. Spool assembly 9 is fastened to flange 5 by means of studs. The spool assembly contains four spools, each stacked of two parts rotating independently. All the four spools are brought to the gear engagement thereby allowing, by initiating one spool, to rotate the rest. The spools are driven with electric motor 10 mounted on flange 5. Axes 11 of the spools are positioned so as to enable rotation in flanges 12 of the spool assembly. The electric motor is powered through collector 1.

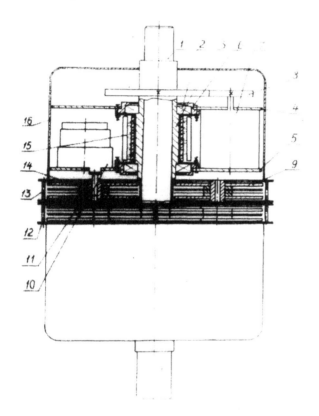

Fig. 6.5. General view of the Proton-5 model design

The entire device is accommodated in casing 16.

### 6.1.3. Experimental Equipment

In conducting experiments on the Proton models the process of building the tether rotating loop was observed by means of the 2TCт32-456 stroboscopic tachometer.

To make revolutions synchronous with flashes of the tachometer lamp the Proton-4 and Proton-5 models were provided with a synchroniser.

The tether configurations have been recorded with cameras Kiev-20 and Kiev-88TT under lighting conditions provided by the stroboscopic tachometer flash lamp.

An illumination power of the stroboscopic tachometer flash lamp was 8 kW.

### 6.1.4. Experimental Results

The experimental results are presented in a sequence as shown in Table 5.

*Table 5*

| Experiment No | System | Number of Experiments | Purpose | Features and Results |
|---|---|---|---|---|
| 1 | CH1 | 17 | Building a circular shape | Tether release velocity − 0.01 m/s with a variable mass of joining members from $10^{-1}$ to 10 g and at the angular velocity from 10 rad/s to 100 rad/s. The tether circular shape has not been obtained. |
| 2 | CH4 | 34 | Building a circular shape of a variable radius | With dependent and independent release velocities of radial and circumferential tethers a specified, stable shape has been obtained. Mass of joining members was derived from Equation 6.1. |
| 3 | CH4 | 10 | Investigation of the case when the radial link is longer than the circular shape radius | Making longer the circumferential tether caused sequential changes of the loop. The system is unstable. |
| 4 | CH | 20 | A process of restoring the circular shape by reducing the length of the circumferential tethers and radial links | To restore the circular shape more time was required against experiments 1−3. Shapes were not repeated in each separate experiment. The system is unstable. |
| 5 | CH4 | 5 | Restoring the circular shape through reducing the length of circumferential tethers | Impossible |
| 6 | CH5 | 26 | Obtaining the circular shape of a changeable radius | The radial links of joining members and circumferential tethers of the loop are released independently. With masses of joining members changed from $7 \cdot 10^{-2}$ to $2 \cdot 10^{0}$ g the direct dependence of the joining member mass from the circumference radius has been established. |

| Experiment No | System | Number of Experiments | Purpose | Features and Results |
|---|---|---|---|---|
| 7 | CH6 | 10 | Obtaining the tether circular shape of a changeable radius. | Impossible |
| 8 | CH7 | 30 | Obtaining the tether circular shape of a changeable radius. | At the initial release of the circumferential tethers with the subsequent release of the radial links and the joining member the circular shape has not been obtained. With the alternative release of the both a circular shape of $D = 1.5$ m has been obtained. The further release resulted in a loss of the circular shape. |
| 9 | CH4 | 10 | Exploring the release of the circumferential tethers only | The circular shape can not be obtained. |
| 10 | HH | | Obtaining the circular shape of a constant radius | The circular shape tolerant to low perturbations was built. With the axis of rotation tilt of 30° no visible changes of the form were observed. |
| 11 | HH | 20 | System stability evaluation with the radial tethers longer than the circular shape radius | Increasing the length of tethers results in a loss of the circular shape stability over $1/5 - 1/3$ of the radius. |
| 12 | HH1 | 5 | Building the tether circular shape of the specified radius | Impossible |
| 13 | CH2 | 15 | Building the tether circular shape of the specified radius | By varying masses of the joining members and angular rotation velocity the set task has not been implemented. |
| 14 | CH3 | 27 | Building the tether circular shape of the specified radius | The joining member mass derived from Equation (6.1) was 1.3 g. The obtained shapes are tolerant to perturbations and tilts of the rotation plane. |
| 15 | CH4 | 20 | Obtaining a system with a zero moment of momentum | Stable systems have been obtained; no engagement between the loops was observed. |

The flexible, fluorineplastic-braded, copper cable tether of a linear density $\mu = 1.925 \cdot 10^{-3}$ kg/m and $d = 5 \cdot 10^{-4}$ m diameter was employed to conduct all the experiments.

Photos have been taken under the synchronised flash lighting. The rotation angular velocity was varying within the range from 10 rad/s to 100 rad/s. At tilts of the plane of rotation the maximum value of the axis deflection from the vertical is $\pm 30$ degrees. The angle is changing at a rate of 0.1 rad/s.

All the experiments demonstrated certain properties common to all systems involving flexible tethers. The air resistance is not radically changing the system configuration, however, it causes the tethers to offset in the direction opposite to the rotation. This poses limitations to the scope of experiments investigating the systems involving flexible tethers in the air environment because the tether deflection exceeding 90 degrees could cause its entanglement during release. Increasing the tether linear density will reduce the deviation angle, and increasing the system dimensions will increase the deviation. With the angular velocity changing within the aforesaid limits and the axis of rotation tilting at an angle of $\pm 30$ no tangible changes of the system configuration have been observed.

Because the investigation of the tethered systems in air is relatively less difficult comparing to the conduction of experiments on film systems in a pressure chamber, a rather good history has been obtained for a large number of options (Table 5).

It is clear from the analysis of experimental results that the system configuration would be stable during deployment only in two cases: the CH4 system with masses of joining members derived from the expression:

$$m = n_c \, n_k \, \mu \, R_k \qquad (6.1)$$

where $m$  — total mass of joining members;
$\quad n_c$  — number of joining members;
$\quad n_k$  — number of tethers fed through the joining member;
$\quad \mu$  — tether linear density;
$\quad R_k$ — system final radius;
and for the HH system with no joining members having a rigid link between the radial and circumferential tethers.

The latter system appeared to be more simple and efficient because of no requirements for masses of joining members being actually equal to the mass of the tethers included in the system, the absence of the sliding contact element in the joining member, and no potential probability of jamming. In Fig. 6.6 photos made under the stroboscopic lamp lighting conditions show how the CH1 system looses its stability. Fig. 6.7 illustrates the steps of building the stable HH system.

In conclusion, it must be noted that the experimental studies of the tethered systems have contributed much to the understanding of how the flexible systems (including the film systems) are deployed. Based on these studies the structures capable to keep stability while being deployed have been selected and further proposed for use in the capacity of flywheels enabling the counter-rotation of film sheets and tethered, loop antennas of the ELF and LF bands. The tethered system deployment mechanisms employed in the laboratory experiments have been accounted to a great extent as the deployment mechanism prototypes incorporated in the follow-on projects of space systems.

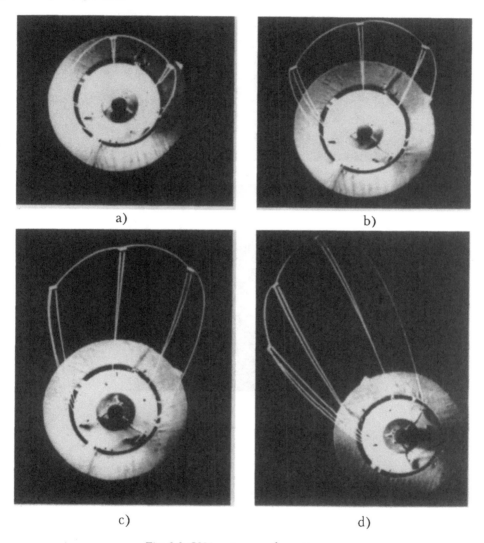

a)

b)

c)

d)

Fig. 6.6. CH1 system configurations

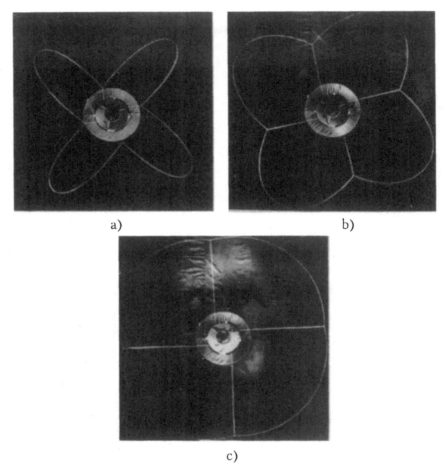

a)                                                      b)

c)

Fig. 6.7. HH system configurations

## 6.2. Ground Experimental Development of Film Sheets in Vacuum Chambers

### 6.2.1. Purposes and Objectives

The first experiments for deploying the film reflectors by centrifugal forces have performed jointly by The Central Research Institute for Machine Building and RSC Energia on the Y-22 vacuum test setup.

The following thin film sheets (disks) were used as the test objects:

• single film sheets of 5 m diameter, both unfolded and folded into two patterns, such as a two-side bellow (Fig. 6.8) and a four-side bellow,

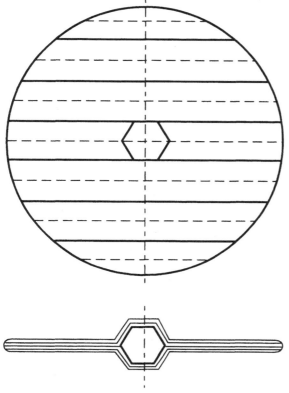

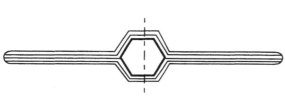

Fig. 6.8. Film sheet folded into the two-side bellows pattern

• single film sheets of 2 m diameter folded into the "six-point star" pattern (Fig 6.9),

•. double ("meniscus-type") sheet of 5 m diameter (two single sheets bonded around the perimeter (Fig. 6.10) with no pre-packaging.

The test purpose was to define principal behaviour characteristics of the rotating film sheets depending on different factors and also to establish the most stable modes for unfolding the film sheets from the stowed pattern.

The objectives pursued by the experimental development were:

a) to define the drive rotation frequency which results in the transition of the film sheet to a plane from a freely suspended state,

b) to establish how the film sheet surface reflectance depends on the drive rotation frequency,

c) to establish how the film surface reflectance and shape depend on the ambient pressure,

d) to explore the possibility of re-pointing the axis of rotation of the film sheet expanded by centrifugal forces,

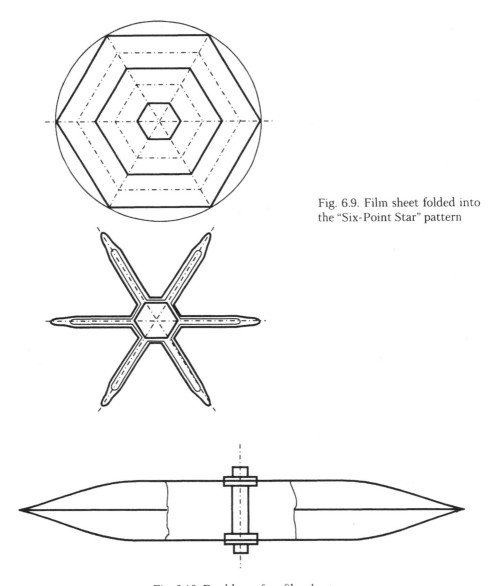

Fig. 6.9. Film sheet folded into the "Six-Point Star" pattern

Fig. 6.10. Double-surface film sheet

e) to define the rotation frequencies which result in destroying the film sheets of different sizes,

f) to identify optimum modes of unfolding the film structures from different folding patterns; to make a comparative assessment of folding patterns,

g) to explore the double film sheets.

The experimental research program has been developed so that to solve several experimental tasks simultaneously.

### 6.2.2. Experimental Setup. Identified Equipment

Experiments have been conducted at the CSRIMB on the Y-22 vacuum setup (Fig. 6.11). The setup operational part is vacuum chamber (1) 8 m in diameter and about 20 m high. The vacuum chamber is provided with vacuum station (9) enabling to generate the $5 \cdot 10^{-4}$ mm Hg vacuum, safety device (12), air supply system (11), and ventilation system (10). The vacuum chamber sides contain special outlets (16) to accommodate leakproof adapters through which the control and measuring system electric utilities are entered. The upper part contains optical panels (3) to allow the accommodation of optical windows of various diameters (800 mm, 400 mm, 230 mm). The middle and lower parts of the vacuum chamber contain vertical struts (17) which accommodate maintenance sites (18) moving around the vacuum chamber inner perimeter.

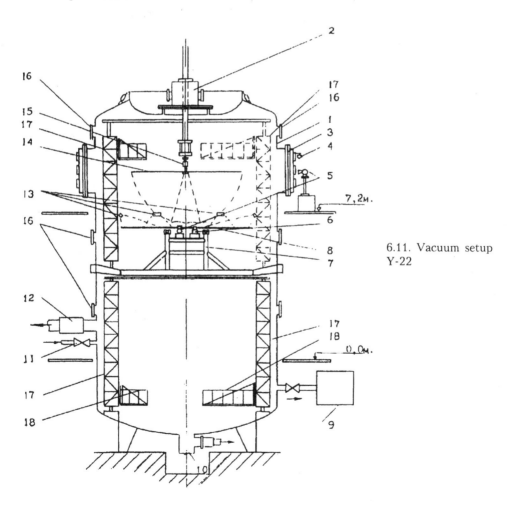

6.11. Vacuum setup Y-22

To conduct the tests, vertical motion mechanism (2) was installed on the vacuum chamber top cover. Experimental electric drive (15) designed and manufactured by the Tashkent Design Office for Machine Building was mounted on the vertical motion mechanism to provide rotation and rotation of film sheet (14).

Light dissipating screen (8) 36 m$^3$ in area made of white cloth was stretched on at the level of support (7). Over the screen (close to its centre), at a height of 0.7 m, test grids (19) were located to investigate the metallized film sheet reflectance by way of taking photos of the grids through the specular surface. Two types of the test grids were used: frames 80 × 200 cm with a system of stretched on rubber cords forming square meshes 10 × 10 cm and wooden boards 150 × 150 cm with the similar grid drawn on them. Eight lights K-1000 (13) were fixed on the vacuum chamber walls. Movie camera AKC-2 (5) and photo camera AHΦ (6) were installed on support (7) to take movies and photos. Camera AHΦ (6) was installed outside facing one of the windows located along the optical line coinciding with the expanded film sheet plane. The Zenith-ET camera was used for the aspect photography outside.

An electric device (Fig. 6.12) was used to spin the film sheets and turn their axis of rotation. This device based on the Д-149 vacuum electric motor, 150 W, has been designed and manufactured by the Tashkent Design Office for Machine Building. The experimental setup consists of motor (1) to spin film sheet (2) fastened to or reeled on spool (3), belt release mechanism (4), and mechanism (5) for turning the axis of rotation. A special feature of mechanism (5) is that the film sheet rotation drive is fastened in a suspender released on an operator command, the turn being performed by the action of the moment of force generated by a weight of 224 g attached to a lever 21 cm long. The motor is provided with the spool rotation velocity pulse sensor.

The experimental film sheets fabricated by the DADO are discs of metallized polymer film Mylar-K-OA 5 m thick composed of strips 600 mm wide laid down next to each other, edge to edge. The seams were reinforced with the heat-activated adhesive tape 20 mm wide. The rotation velocities estimated for the film sheets are shown in Table 6.

Fig. 6.12. Film structure rotation electromechanical device designed by the Tashkent Design Office for Machine Building

*Table 6*

| Diameter, m | Operational velocity, rev/min | Critical velocity, rev/min |
|---|---|---|
| 2 | 743 − 1050 | 3322 |
| 5 | 297 − 420 | 1328 |

Prior to perform the experiments the experimental equipment was mounted and control and measuring circuits were hooked up. Then the spool with the freely suspended or folded film sheet was installed on the drive. Sheets of 5 m diameter were folded on a free, smooth area of the facility. A sheet of 2 m diameter was folded into the production pattern ("the six-point star pattern") at the DADO. Thereafter the vacuum chamber was evacuated to $10^{-3}$–$10^{-2}$ mm Hg. During the experiments involving the freely suspended film sheets an operator, controlling the motor voltage manually, set the film sheet in rotation. On reaching the required rotation frequency the film-photo cameras started to record the rotating film sheet. Simultaneously, indications of instruments were read-out.

When analysing the pressure impact on the film sheet shape, the pressure was increased in steps starting from $10^{-3}$ mm Hg, the film sheet being photographed and the drive parameters being recorded.

In performing experiments for unfolding a film sheet from different patterns the spool had been spun to a specified velocity, whereupon the cable belt was thrown off the spool on an operator command and the film sheet was unfolded (with the simultaneous film-recording).

To study the process of turning the film sheet axis of rotation, the film sheet had been spun to 500–600 rev/min and then, on an operator command, the rotation drive was released in the suspender. Thereafter, the electric drive and the rotating film sheet deflected from the vertical at an angle of about 20 degrees under the moment of force generated under the action of a levered weight. The process was recorded by the film-photo cameras.


*6.2.3. Experimental Results*

In performing the experimental studies on the Y-22 setup the following results have been obtained:

a) it had been established that the electric drive rotation frequency, whereby the freely suspended film sheet was transferred into a plane, was being maintained through all the experiments at a level of about 100 rev/min; Figs. 6.13 a–c show steps in which the film sheet is transferred from a freely suspended state to a plane,

b) the specular surface has been formed on the rotating film sheet of 5 m diameter at frequencies in excess of 400 rev/m, the smallest distortions being observed at 600–700 rev/min; it must be noted that in the central area, where the sheet was attached to the spool (0.5–0.7 m), small distortions were observed likely caused by tensions generated in the attachment area owing to the force of gravity and vibrations of the driving shaft; while testing sheets of 2 m diameter in a similar way, the

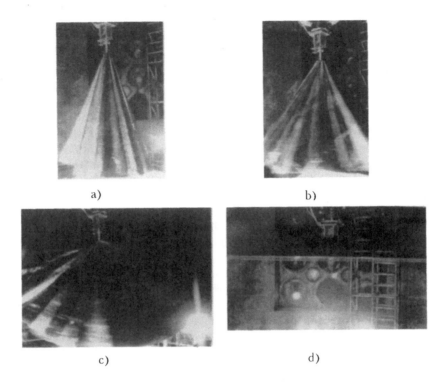

a)    b)

c)    d)

Fig. 6.13. Steps of the film sheet transfer into a plane from the freely suspended state

distortions were observed over the entire surface; the most contrast image of the test grids being photographed through the specular surface has been obtained at the electric drive rotation velocity of 1000 rev/min; the further increase of the rotation velocity resulted in a drastically deteriorated specular surface due to the visually observed oscillations of the electric drive; Fig. 6.14 shows the aspect picture of the expanded film reflector of 5 m diameter with a specular surface; the picture shown in Fig. 6.15 was taken by photographing the test grid through the 5 m reflector at a rotation velocity of about 600 rev/min thereby demonstrating a good quality of the specular surface,

c) studies on how the pressure within a spectrum from $10^{-3}$ to 0.5 mm Hg impacts the rotating film sheet indicated that the specular surface quality would deteriorate as the pressure increased; at a pressure exceeding 0.5 mm Hg the film sheet would cease to be flat and turn into various space configurations (Figs. 6.16 a,b) and, as a consequence, would get into a tangle,

d) in performing studies on how the axis of rotation is turning, the film sheet, while being rotated at a velocity of 570 rev/min, is turned 20 degrees at an angular velocity of about 5 deg/s; in the process of turning a certain deterioration of the specular surface quality is observed, however, 2—3 s after the turn completion, this effect disappeared with the reflection characteristics recovered,

Fig. 6.14. Deployed sheet of $D$ = 5 m with the specular surface

Fig. 6.15. Reflection in the test grid

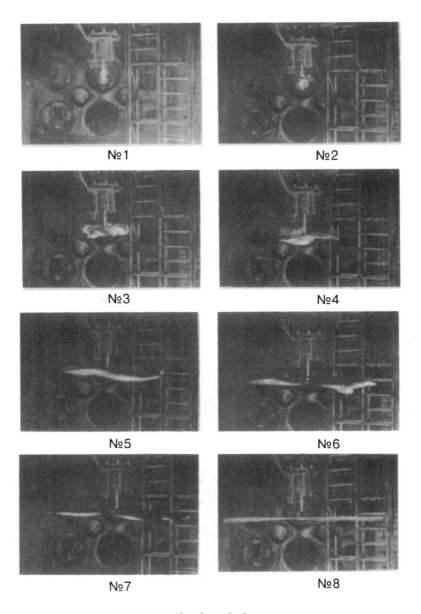

Fig. 6.16. Film sheet deployment steps

e) the double "meniscus-type" film sheet has been designed as two single strips, bonded around the perimeter and spaced in the central area at 19 cm; this sheet has been deployed into the designed shape at the rotation velocity of 120—200 rev/min and demonstrated its reflectance at 300 rev/min; however, at 380 rev/min the film fracture was observed.

Of particular emphasis is a set of experiments for unfolding the film sheets from different folding patterns.

Two efforts undertaken to unfold the sheets of 5 m diameter from the two-side bellows pattern commenced from a rotation velocity of 1100 rev/min. While deploying the sheets, the rotation velocity decreased to 60 rev/min thus making the sheets to drop and get into a tangle. It must be noted that the above pattern tended to be deployed not completely, mainly, in the longitudinal direction of the bellow-type, i.e. the folded strips of the film sheet would retain their shape during deployment and would not stretch laterally.

The 5 m film sheet folded into the "four-side accordion" pattern has been deployed with the initial velocity of 1400 rev/min. In the process of deployment the rotation velocity decreased to 120 rev/min. The film sheet has been partially deployed, the deployed area being 80% of the total sheet area; its central portion looked like a short rope commencing at the spool. This effect could be attributable to that the deployment was non-uniform in all directions. It is noteworthy that by comparing the processes used to feed out the sheets from the four-side and two-side patterns the undoubted advantage of the four-side folding pattern has been recognised.

Good results have been obtained from deploying the 2 m film sheet folded into the "six-point star" pattern. The deployment initiated at 1200 rev/min proceeded uniformly in all directions completing at 480 rev/min; the deployment time was 1 s that was confirmed by the motion-picture recording (Fig. 6.16 a,b).

The experimental equipment provided by the Tashkent Design Office for Machine Building has appeared to be inadequate to enable the full scope of studies on film structures built by centrifugal forces.

There, the rotation drive was insufficiently powerful to test film sheets of 5m diameter at rotation velocities ranging from 800 rev/min to the film fracture velocity. The same reason prevented from testing the double sheets in a full scope.

The device turning the axis of rotation was insufficiently rigid while being secured and thus tangibly affected the process of investigating the specular surface performance at large rotation velocities due to strong oscillations.

A poor location of the belt release mechanism could be also applied to a design drawback that, in certain cases, resulted in the film sheet damage during the experimental deployment.

In this connection, a necessity emerged to continue investigations on the film reflector models using the updated experimental facilities.

The ground experimental research in the structureless film reflectors expanded by centrifugal forces has been performed further in 1990 at RSC Energia in the ЭН-85-16 vacuum chamber.

The experimental program, as a whole, was pursuing the same goals as the research program previously implemented by the CRIMB; the experimental setup is schematically shown in Fig. 6.17. However, the two sets of experiments exhibited a number of sufficient distinctions.

First, to spin the film sheets RSC Energia employed a powerful electric drive designed and manufactured by the Moscow Aviation Institute. The electric drive (Fig. 6.18) consists of: spinning drive derived from a commercial electric motor

RSC Energia experimental setup diagram

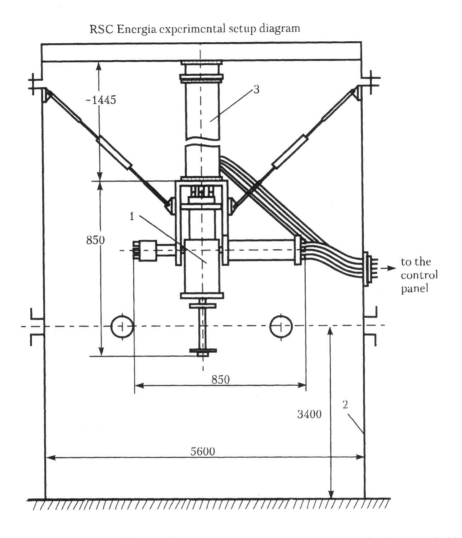

Fig. 6.17. Test setup installation scheme. 1 — setup ИГЭ-1; 2 — vacuum chamber; 3 — holder

МСШ2М, 1 kW, and housed in a leak-proof case (1); a rotation axis turn motor-re-ducer housed in a leak-proof case (2); piezo-optic sensor (3) for measuring the rota-tion velocity of spool (7); mechanism for unpinning the folded film sheets (4); the turn motor-reducer electromagnetic brake coupling (5) and associated rotation an-gle sensor (6). Leak-proof case (1) is accommodated in suspended frame (8), leak-proof case (2) is fastened to load-bearing frame (9).

Such an electric drive enabled the film sheets to be investigated in a wide spec-trum of rotation velocities up to the breaking rotation velocity (about 1330 rev/min).

Fig. 6.18. Film sheet rotation electric drive of the Moscow Aviation Institute design

Second, unlike the experiments conducted at the CRIMB, where the axis of rotation were being turned in a free suspender, in the set of experiments under consideration the motor-reducer was performing this function maintaining a constant angular velocity of rotation. This brought up the question of how to determine parameters of the running wave and a damping decrement of oscillations in the film sheet caused by turning the rotation axis at a constant angular velocity.

Third, the investigations have been conducted not only on different folding patterns but also on different shapes of film surface.

The experiment process flow has been maintained, as a whole. It could be added that, while investigating the film sheet rotation axis turn and determining the oscillation damping decrement, the film sheet rotation velocity was brought to 1000 rev/min, and then the spinning electric drive was turned by means of the motor-reducer. There, the rotation axis was deflected 15° from the vertical at an angular velocity of 1 rev/min. Upon attenuation of the film sheet, the original vertical position has been restored through the motor-reducer reverse run. The film sheet oscillation amplitude, shape, and oscillation processes were recorded by the movie cameras. The time of the running wave complete decay has been also recorded.

The film sheet has been deployed from the two-side bellow pattern through the manual control of the drive rotation velocity. The deployment had been initiated at 100 rev/min. The film sheet started to be fed out non-uniformly and a tendency to a one-sided deployment of the film disc was observed. As a consequence, the film sheet had to be shaken up by the motor shutdown and the follow-on smooth rotation at a velocity much the same as the sheet rotation inertial velocity. Only after the sheet had been shaken up three times, it restored the disc shape.

In the second experiment a film sheet folded into the two-side bellows pattern has been also used. The deployment had been commenced at 200 rev/min and proceeded with no operator involved. At the third second from the deployment initiation the film sheet was built into a horizontal, folded in two disc with the rotation velocity reduced to 140 rev/min. 7 seconds following the deployment the film sheet broke away from the spool in the attachment point that likely resulted from notching the film with metal rims of the spool.

In the third experiment the pattern as per ref. [2], different in a rotation symmetry and a radial structure of folds, has been used. (See Fig. 3.2). From theoretical considerations this pattern offers notable advantages in deployment as compared to the bellow-type patterns because the rotation symmetry and a large number of folds afford the uniform deployment of the film disc in all directions, and, upon the deployment completion, no quick impact is suffered by the film material.

However, the deployment from such a pattern was not a success, because the structural compexity of folds and durable stowage in a closely packed pattern caused the film material to be hardly engaged inside the package thereby interfering with the deployment process. Upon opening the vacuum chamber, the folds and engagements had been smoothed out and the experiment was repeated with the cantilevered film sheet. In testing this film sheet its critical rotation velocity has been determined. This velocity was gradually decreasing and, on reaching 1400 rev/min, the film sheet was destroyed. Practically, this is in agreement with the theoretical estimations predicting that the film sheet of 5 m diameter would be destroyed at 1330 rev/min.

To investigate the process of turning the film sheet rotation axis, two types of sheets with differently shaped surfaces have been employed.

The rotation axis turn has been initiated at 1000 rev/min within a range of 15 degrees. While turning the axis, the film disc precession was observed followed by the surface oscillations. Principal characteristics of the rotation axis turn process are given in Table 7, where $T_0$ — period of precession, s; $A$ — maximum amplitude of disc edge oscillations, cm; $A_0$ — amplitude of disc oscillations in a steady state, cm; $t$ — oscillation damping time, s; $\delta$ — oscillation damping decrement derived from the Expression:

$$\delta = \ln \frac{A_i}{A_{i+1}}$$

*Table 7*

| Film N | $T_0$ | $t$ | $A$ | $A_0$ | $\delta$ |
|--------|-------|-----|-----|-------|----------|
| HM0002-0 | 60 | 600 | 30 | 1.5 | $5.0 \cdot 10^{-4}$ |
| HM0003-02 | 50 | 450 | 20 | 0.3 | $5.8 \cdot 10^{-4}$ |

By comparing the two types of the film patterns, a conclusion could be made saying that the pattern only slightly affects the dynamics of rotation and re-orientation of the film sheets.

It was remarked that the lower is the angular velocity at the completion of the turn, the smaller is the maximum amplitude of the film disc edge oscillation.

In the experiment using the "meniscus-type" sheet with the doubled surface the sheet was observed to take an operational configuration at a rotation velocity of 300 rev/min. At 1000 rev/min an attempt made to turn the film sheet rotation axis resulted in the film sheet failure likely because of a stress occurring in the film material during the turn. Due to the fact that the "meniscus-type" sheet is of twice as much mass and has a more rigid structure comparing to the single film disc, strong stresses would be inevitably generated such a sheet.

The acceleration (boost) stage of the completely deployed film reflector requires a particular consideration. Rigorous numerical calculations have not been made in this area because of the possible waviness on the film sheet in the area where it is attached to the spool. The waviness is associated with tangential stresses generated during the film sheet acceleration. The possible occurrence of waviness (compressive stresses) on the soft skin would make the task sufficiently different from the traditional task dealing with the rigid disc acceleration.

The possibility has been considered to accelerate the film sheet by applying the well-known theory for the rigid disc with no compressive stresses.

The following expression has been obtained for a relative acceleration of the disc:

$$\bar{\varepsilon} = \frac{\dot{\omega}}{\omega^2} = \frac{\nu + 3}{\nu + 1} \, \nu^{1/2} \, \eta_0{}^2$$

where:

$$\eta_0 = \frac{R_0}{2 R_k}$$

The acceleration time is determined as:

$$t = \frac{2(1 - \alpha)}{\omega_{max} \, \overline{\varepsilon} \, \alpha}$$

where:

$$\alpha = \frac{\omega_{min}}{\omega_{max}}$$

For $R = 0.15$ m; $R = 10$ m; $\nu = 0.4$ — the Poisson's ratio; $\alpha = 0.5$; $\omega = 7.7$ rad/s; $\overline{\varepsilon} = 8.10^{-5}$; $t = 50$ min.

The time, $t = 50$ min, did not meet the requirements of the Znamya-2 experiment program. A question arose of applying this expression and substantiating the possibility to reduce the acceleration period. The experimental results (par. 6.2) show the possibility to reduce the acceleration time more than by an order of magnitude.

Let us analyse the experimental results obtained while rotating the film sheet in the vacuum chamber. During the experiments the acceleration has not been measured directly. However, due to relatively poor vacuum ($10^{-1}$ mm Hg) and the associated gas-dynamic resistance of the film disc, a considerable power consumption of the electric drive was observed at constant angular rotation velocities. For example, with the sheet $R = 2.5$ m, $R_0 = 0.03$ m, $\omega = 700$ rev/min at $P = 120$ W.

It has been demonstrated that the resistance in bearings is negligibly low. Hence, the gas-dynamic resistance caused a torque on the motor shaft and, within the area where the disc was attached to the electric motor drive shaft, this torque created a combined stressed state equal to the action of certain effective acceleration. Actually, from the equation of momentum we have:

$$J \dot{\omega} = M \, ,$$

where $J = 0.5 \pi \mu R^4$ — moment of inertia of the disc;
$\mu = 6.5 \cdot 10^{-3}$ kg/m$^2$ — surface density of the film.

Multiplying both parts of the —quation by angular velocity we obtain the equation of power:

$$J \dot{\omega} \omega = \omega M = P \, ,$$

from which we derive:

$$\dot{\omega} = \frac{P}{J \omega}$$

For the aforesaid mode we have:

$$J = 4 \cdot 10^{-1} \text{ kg/m}^2; \; \dot{\omega} = 4 \text{ rad/s}^2; \overline{\varepsilon} = 10^{-3}$$

There, a theoretical value of relative acceleration is :

$$\overline{\varepsilon}_{theor} = 5 \cdot 10^{-5}$$

Hence, a ratio of the experimental to relative accelerations is:

$$\frac{\overline{\varepsilon}_{exp}}{\overline{\varepsilon}_{theor}} = 20$$

In performing the experiments the film sheet has not been destroyed because of the inadequately powerful electric motor (par. 6.2) and hence it appears that the experimental value of the effective acceleration is not a limit.

The further research in the flexible disc acceleration process was based on the multi-rod model loosing its stability, zone by zone, in the process of acceleration. A restriction for the relative acceleration has been established, viz

$$\overline{\varepsilon} < 5 \cdot 10^{-3} \dots 10^{-2}$$

which is proved by the foregoing experimental data.

For the acceleration mode in the Znamya-2 experiment

$$\overline{\varepsilon} = 10^{-3},$$

that is applicable to the given range.

Physically, the expression for the relative acceleration, $\overline{\varepsilon} = 10^{-3} = \dot{\omega}/\omega^2$, is a ratio of the inertial to centrifugal forces and, at low values of the ratio, an angle of fold deflection from the radial direction.

The above facts could serve as a substantiation for the acceleration predicted for space systems.


## 6.3. Conclusions

Experimental research of film reflectors expanded by centrifugal forces, which has been conducted within the period of 1986 through 1990, are of a great importance in defining the space reflector configuration and dynamic parameters.

During the given experiments the film reflector expanded by centrifugal forces has been originally deployed under the ground conditions thereby confirming the possibility of deploying a similar large reflector in space, the more especially as no unfavourable factors, inevitable during the ground tests, exist in space (such as gravity, residual atmospheric pressure).

The rotating metallized film disc has been demonstrated as capable to create a specular surface. The possibility has been shown to change the specular surface curvature by using the double ("meniscus-type") film discs. In this way the applicability of such film structures in the capacity of electromagnetic radiation retro-reflectors in a wide spectrum of wavelengths has been proved.

The experimental evidence points to the fact that the structures being verified could be re-oriented in space, the capability being of a great importance as, otherwise, the structureless film reflectors expanded by centrifugal forces would not be promising for the specific application. It must be emphasised that, during the two experimental sets, two radically different dynamic modes of re-orientation have been revealed, each of the modes using a different way of turning the film reflector rotation axis.

As it is apparent now, the researchers, while conducting the experiments at the CRIMB, have inadvertently entered a zone of passive damping. This resulted from the fact that the suspender of the experimental electric drive rotating the film sheet was not rigid and was of a low mass. By accident, a feedback had occurred and completely damped the possible oscillations. When the experiment had been repeated at RSC Energia employing a heavy, hard-fastened electric drive, a spectrum of oscillations was obtained and a necessity of their damping was demonstrated.

Experimentally, the possibility to increase the acceleration of the fully deployed film sheet beyond one order of magnitude versus the estimated value has been proved.

A set of experiments on the deployment of film sheets from folding patterns revealed the undoubted advantages offered by the multi-beam folding patterns having a radial structure. Further, this conclusion appeared to be a basis in defining a folding pattern for the reflector employed in the space-based experiment and in calculating the deployment dynamics of the film structure.

The experiments were accompanied by solving a number of methodical and technological tasks such as: sheet acceleration rate, impact of vacuum level, effects of the film sheet thickness, structure, tension when folded as being fabricated, time of stowage, thickness of stick-protection talcum powder covering. The experience gained from the experiments was further incorporated in programs-methods used for ground verification of large systems.

# CHAPTER VII

# DEVELOPMENT OF SPACE-BASED FILM REFLECTOR OF 20 m DIAMETER (ZNAMYA-2 EXPERIMENT)

## 7.1. Developmental Concepts

To prove design concepts and computing techniques, the space-based Znamya-2 experiment has been prepared and accomplished on 02/04/93.

The development was based on the following concepts:

• using the Progress controllable cargo vehicle in the capacity of a scientific hardware carrier and its unique capabilities to accommodate various space-based experiments, as well as technological and recording capabilities of Space Station Mir,

• incorporating at the most the deployment mechanism assemblies and units previously used in the Progress vehicle structure; involving specialists and teams previously developing and operating these structure elements,

• employing at the most the experience and techniques derived from accommodating other space experiments on the cargo vehicle and Space Station Mir,

• making the maximum simple and reliable solutions to the tasks having no analogues in the world practice.

The last was applied to the selection of a folding pattern, film sheet deployment dynamics, and active damping of the film sheet surface oscillations during the vehicle turn.

## 7.2. Reflector Folding Patterns and Deployment Dynamics

Based on a large number of ground experiments conducted in vacuum chambers at RSC Energia and CRIMB a conclusion was reached that the deployment would not be stable with the film being fed out from simple prima facie folding patterns (of the bellow-type) and a necessity was revealed of using the multi-beam radial folding patterns possessing, due to the multi-beam structure, the symmetry in the field of centrifugal forces while being deployed.

The film sheet deployment dynamics turned to be a key task in the given experiment. Ultimately, the deployment dynamics defined the deployment mechanism structure, teams of specialists to be involved in the experiment, the experiment outcome, and outlooks.

In the Znamya-2 experiment this task was being solved in the following way: a priori, several equations have been derived for the first time (Chapter 3) to describe the film reflector deployment dynamics in the field of centrifugal forces. The equations contained angular velocities and rotation accelerations, angles of deflection from the radial direction, their derivatives, forces over the structure radius associated with a moment of force acting on the system. The possibility of applying the moment from various existing hardware items such as a flywheel accumulator, a powder or gas nozzle, electric drives with rigid and drooping characteristics. Different laws describing how the moment and angular velocity are change in time have been considered. By the numeric solution of the task unstable results have been received in all cases except for the case when an electric drive with a soft (drooping) characteristic is used. It is known that such a characteristic is offered by all d.c. motors which are widely used including the Progress docking unit.

Based on the Progress "probe and drogue" docking unit the configuration of the reflector deployment mechanism, i. e. mounting points, dimensions, power supply cabling, and electric drives have been defined.

The reflector sheet size had been determined from the docking unit dimensions and was varying from 20 m to 25 m within the period of the hardware development. In view of the fact that the experiment had been prepared to prove the design solutions, dynamics of deployment, and space attitude control, it was undesirable to build a huge reflector because of its high cost.

The point to be made again is that the stable deployment of the film sheet provided at the expense of the d.c. motor drooping characteristic, $M = f(w)$, has made the design successful, reliable, and capable to undergo the ground verification within a short time frame. With other possible motors, for example, a gas motor, the development, testing, and verification of the systems would be complicated and a staff of executors would be different.

## 7.3. Film Sheet Structure

The film disc has been constructed of 8 separate sectors put together around the periphery (Fig. 5.9). An outer diameter was 20 m, an inner diameter — 2 m. The reflector material was the Mylar film 5 µm thick, metallized with aluminium on one side. To reinforce the structure, the aramide cord was built around each of the sectors. Reinforcing members and joining loops were provided in corners.

Each of the sectors was folded into the bellow-type pattern having radial generatrices and reeled up on a spool 160 mm high. The inside edges of each sector were provided with four end tethers 1.25 m long, fastened in pairs to the spool rims (Fig. 7.1).

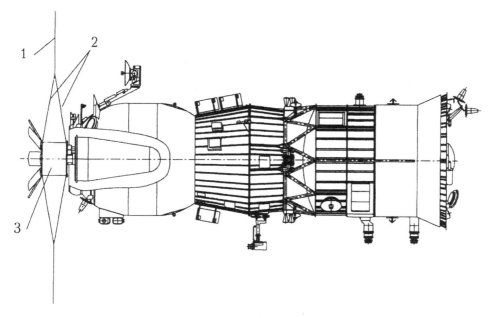

Fig. 7.1. Progress-accommodated reflector layout. 1 — reflector film surface; 2 — guying wires; 3 — central cylinder containing spools for reflector reeling up

In this way the guying-type scheme has been obtained to attach the film sheet to the driving centre. The guying tethers made an angle of 4 degrees to the radial direction. It was expected that, during the vehicle turns, this attachment scheme would save as a rigid insert at angles of the sheet deflection from the normal to the axis of rotation up to 4 degrees. Rims of the spools were allowed to rotate freely around the axes thereby solving the problem of how to distribute layers, by their number, of the reflector sectors on different spools in the process of deployment. The film sectors have been made and reeled up on spools under tension at the DADO. At RSC Energia the spools have been assembled into the reflector deployment mechanism. Based on the experience gained from testing the film structures on vehicles designed by

RSC Energia, the Mylar film reflector service life in open space has been defined as 2 days.

Based on thermal calculations a temperature increase was expected on the film sheet (of the order of 60—80°C) impairing strength characteristics of the structure. This is why the lighting conditions for the TV-photo recording and the film sheet heating conditions had to be compromised.

In the Znamya-2 experiment the film sheet plane was pointed at an angle of 34° to the Sun. The deployed reflector rotated at an angular velocity of 1.8 rad/s, the maximum design stresses at the root cross-section of the film sheet being $\sigma = 4 \, 10^6 \, \text{N/m}^2$.

## 7.4. Reflector Deployment Mechanism Design

The reflector deployment mechanism dimensions were such as to match mounting points on the Progress vehicle probe-and-drogue docking unit. Its mass was 40 kg. To accommodate the deployment mechanism the docking unit mounting frame and cabling were used. The reflector deployment mechanism contained a central hinged beam capable of tilting within an angle of 360° at the expense of the crosswise-mounted electric rolling drives. On the central beam the rolling bearings were provided to mount the central rotating cylinder containing 8 spools combined via the worm gearing deployment drive.

In the process of deployment the spools were rotating at a steady angular velocity of the order of 1 rev/s. The deployment velocity was variable because the sheet strip diameter was varying as the film was fed out from the spool. The total time of deployment was 200 seconds. The central cylinder containing the spools was rotated by two electric drives having different slopes characteristics including a high-velocity drive providing the cylinder initial rotation at up to 10 rad/s. Further, in the process of deployment the angular velocity was reduced to 1 rad/s and the deployment was continued at the low velocity of the second electric drive.

Once the film sheet had been deployed that was recorded by a rotational velocity sensor, the high velocity drive was automatically disconnected and the deployed sheet was accelerated in accordance with the low-velocity drive performance coming to a no-load velocity of the order of 1.5 rad/s. This mode has been maintained within entire experiment up to the deployment mechanism jettison.

In a transport configuration and at the stage of the initial acceleration of the central cylinder at a velocity of 10 rad/s the spools were covered with 8 flaps. While performing handling operations, the central cylinder and spools were additionally covered with a protective sheath. 20 seconds after the spools had been set in rotation by the high-velocity drive, the flaps were simultaneously and automatically opened.

The reflector deployment mechanism has been manufactured by the Instrumentation Enterprise at RSC Energia and verified through the laboratory and qualification tests, as well as through simulating, by means of a tethered weight system, a dynamic load acting on electrical and mechanical performance of the drives, a dynamic load acting on performance of the electric drives (see Chapter 3, par 3.9). There, the film sheet of 4 kg in mass was replaced with two weights of 2 kg each. In the vacuum chamber the weights were hanged from a rotating frame 8 m high and fed out 1.75 m long.

## 7.5. Features of Experiment Program

On February 4, 1993, at 3:42 Moscow time, the Znamya-2 experiment has been initiated and the complete experimental program has been accomplished with no failures and contingencies. The experiment was accommodated on the Progress 215 cargo vehicle being a priory equipped with scientific hardware for conduction of the Znamya-2 experiment. The scientific hardware included such items as the reflector deployment mechanism with the reflector reeled up on spools, the reflector deployment mechanism control system, two exterior TV-cameras for the downlink broadcasting of data on the experiment. The Progress 215 has been lofted to orbit on October 26, 1992 from the Baikonur launch site and docked to Space Station Mir on October 29, 1992.

For performance of the experiment the reflector deployment mechanism was installed in the mounting point of the probe-and-drogue docking unit by the Mir crew members 3 days prior to initiate the experiment. The crew members had performed final electric tests before the vehicle was separated from the station and recorded the experimental hardware installation steps using the TV and photo cameras.

In developing the Znamya-2 experiment program and procedure the Progress attitude with its longitudinal axis being approximately normal to the orbit plane and the film reflector located in the vicinity of this plane was taken as the vehicle reference orientation to be maintained through the experiment. This was dictated by a desire to avoid at the most the drag impact on the reflector.

To optimise conditions for recording the reflector deployment steps including other events of the experiment by the TV and photo cameras, as well as to maintain an optimal temperature mode of the reflector material, an angle between the reflector plane and its direction to the Sun had to be 30—50° (about 34° as on 02.04.93).

All steps of the experiment have been implemented in the sun-lighted part of the orbit.

In conducting the Earth illumination experiment Novi Svet (New Light) the Progress vehicle orientation was based on the condition that the reflected beam would approximately coincide with the undersatellite point while flying over the terminator. As this took place, the crew had to observe the reflected sunlight spot below the Space Station, on the night surface of the Earth.

The retro-reflected sunlight spot motion trajectory was coming over the night surface of the Earth travelling through Lyons, Bern, Munich, Prague, Lodz, Brest, Gomel.

To observe and record the experiment steps, such equipment as video camera LIV, photocamera Hasselblud, and surveying instrument Neva were accommodated on Space Station Mir. In addition, two TV-cameras КЛ-140-СТ-П were provided on Progress.

The experiment steps were directly broadcasted by the Progress TV-cameras. Recording from Mir was performed in parallel with the videotape recording followed by the downlink to the ground stations through the relay satellite.

With a view to receive the most complete data on the reflector deployment process and to support the one-day timeline of the experiment, a decision was made to separate the Progress vehicle and Space Station Mir on the 16-day orbit on February 4, 1993.

The Znamya-2 experiment lasted within the 16 3-day orbits on February 4, 1993.

16-day orbit — separation of the vehicle and the station,
   backout of the vehicle 160 m away from the station in 9 min and occupying a position in front of illuminator N9, axis $X$ of the vehicle being 8° from the normal to the orbit and axis $Y$ — opined at 30° to the Sun,
   — the reflector deployment in 200 seconds (Fig. 7.2).

1-day orbit — the Progress vehicle turns together with the reflector at an angle of 3° and at an angular velocity of 0.2 deg/s.

2-day orbit — the Progress vehicle turns together with the reflector at an angle of 112° and at an angular velocity of 0.2 deg/s (to provide orientation required for the Novi Svet experiment) (Fig. 7.3).

3-day orbit — Novi Svet experiment,
   — the reflector axis tilting at an angle of 3° and at an angular velocity of 0.4 deg/s by means of the axis deflection mechanism,
   — damping of the resulting oscillations,
   — jettisoning the Znamya-2 scientific hardware, completion of the experiment.

All steps of the experiment have been conducted within the visibility zones of ground measuring stations.

The successful accomplishment of the Znamya-2 experiment has been reported by the world media.

## 7.6. Theoretical Concept Versus Experimental Results

Let us consider the applicability of the proposed engineering concept describing how to deploy space-based structures by centrifugal forces from a stowed pattern and re-orient them in space.

One of the Znamya-2 experiment goals was to verify techniques employed to calculate dynamic parameters of deployment from a stowed pattern and re-orientation of the film reflector built in space by centrifugal forces. The complete experiment

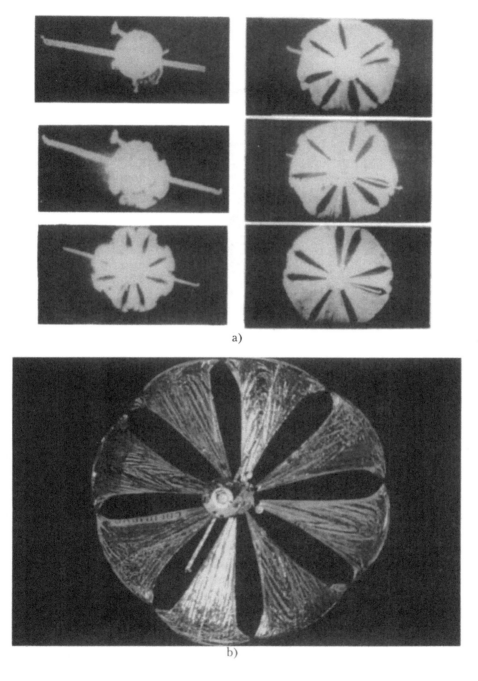

Fig. 7.2. TV-photo recording of Znamya-2 refelector during space-based experiment. a) reflector deployment; b) deployed reflector (photo taken from Space Station Mir with camera Hasselblat); space Station Mir and Progress-M vehicle are about 400 m apart.

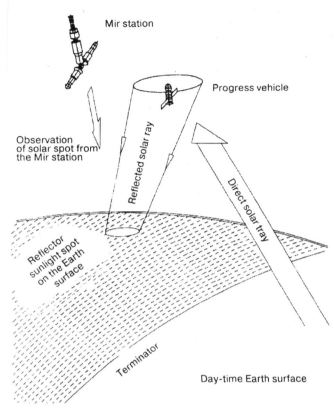

Mir station

Progress vehicle

Observation
of solar spot from
the Mir station

Reflected solar ray

Direct solar ray

Reflector
sunlight spot
on the Earth
surface

Terminator

Day-time Earth surface

Fig. 7.3. Experiment on the
Earth illumination with reflected sunlight during the
hight hours

program has been successfully implemented on February 4, 1993 on the Progress-M
cargo vehicle N215 with no contingencies.

The film reflector 20 m in diameter, composed of 8 separate sectors combined
around the periphery, has been deployed in the vicinity of Space Station Mir. The
video and telemetry data was transmitted to the ground stations enabling to establish, in the process of deployment, time dependencies of such important parameters
as angular velocity $\omega$ , angle of deflection from radial direction $\varphi$ , the reflector actual
radius $R_i$, and an angle of the reflector edge deflection from the rotation plane during
the reflector re-orientation in space.

Records of angular velocity  has been made at a high accuracy for a period of signals from the angular velocity sensors located along axes $Y$ and $Z$ of the Progress-M
vehicle and responding to a small unbalance of the film reflector following the deployment initiation. The angular velocity potentiometer incorporated in a telemetry
system provided a sufficiently lower accuracy of measurements.

The reflector deployment video records made by cosmonauts using the LIV video
system through illuminator N9 allowed to define an angle of deflection from the radial direction of the central line of sectors. A freeze frame clearly shows the sector
centre line deflection angle from the plane of easily observable, radially pointed solar
arrays. In the process of the reflector deployment the Progress-M vehicle was oriented so as to enable the video camera to record the Progress-M aft and the solar ar-

rays in the rear. The solar arrays were spanned over 10 m thereby specifying a size for comparing with the reflector being deployed on the Progress-M docking unit and observed in the background of the frame. Within the 200-second deployment of the reflector the vehicle was continuously moving away from the station and the cosmonaut-operator was periodically adjusting the definition of the image used to make the real time records on the angle of deflection and the actual radius relative to the size of the solar arrays.

The other important factor being recorded in parallel was the lack of oscillations between separate sectors within the plane of rotation and the rectiliniarity of the film folds on the sectors (no falci-form). Fig. 7.4 shows the experimental data and calculated dependencies received under the foregoing techniques (Chapter 3).

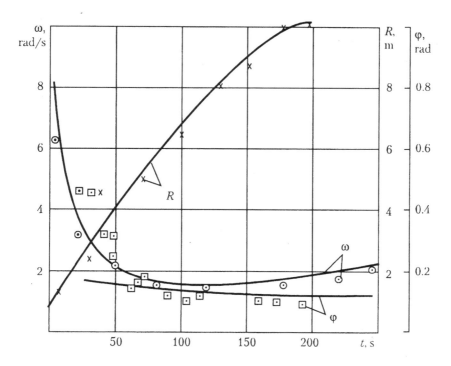

Fig. 7.4. Film sheet deployment process. $\omega$ — angular velocity, rad/s; $\varphi$ — deflection angle, rad; $R$ — current radius, m; $\odot$ — telemetry from the angular velocity sensors; $\odot$ x — wide-recording; —— —theoretical calculations

In the Znamya-2 experiment the angle of the film sheet deflection from the plane of rotation caused by the rotating sheet precession during re-orientation in space was measured through processing the video records taken by the TV camera accommodated on the Progress-M vehicle so that to record the deflection of the film sheet edges.

Simultaneously, telemetry data were recorded from the angular velocity sensors located along the Progress-M $X$ and $Y$ axes. The sensors responded to the amplitude, period, and direction of a travelling wave associated with the reflector precession.

The two-step angular turn of the Progress-M vehicle has been assumed for the experiment with the film sheet being rotated (angular velocity of rotation, $\omega$, following the data from the angular velocity sensors, run to 1.82 rad/s) at angular velocities coming to $\Omega_1 = 0.2$ deg/s and $\Omega_2 = 0.4$ deg/s during re-orientation.

In designing the film reflector and preparing the space-based experiment the reflector angle of deflection, $\alpha$, from the plane of rotation was calculated from the following expression derived from the equation of constrained oscillations of the tethered weight system:

$$\alpha = \frac{2\,\Omega}{\omega}\frac{R_k}{R_0} \qquad (7.1)$$

where $R_k$ — actual radius of the reflector ($R_k = 10$ m);

$R_0$ — radius of the central guying tether ($R_0 = 1.25$ m).

Fig. 7.5 shows the experimental data on angle obtained by processing the TV images and telemetry data. A solid line is marked on the expression (7.1). The reverse-travelling wave period came to 40 s.

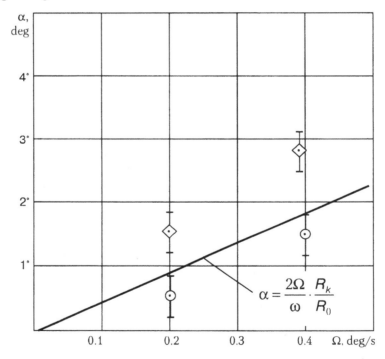

Fig. 7.5. Maximum angular amplitude of the sheet edge deflection at re-orientation in space.
◇ — telemetry from the angular velocity sensors;  ⊙ — video-recording

While preparing for the Znamya-2 experiment, the deployment mechanism ground verification has been performed including the dynamic simulation of loads acting on the film reflector deployment mechanism by means of tethered weights.

There, based on the variants numeric calculation by the techniques applied to the tethered weight system, masses for 8 weights have been established at an accuracy sufficient for design values of tension in root cross-section N, deviation angle $\varphi$, and a range in which the angular velocity $\omega$ is varying (within a shorter period of time) during the model system deployment in a limited volume of the vacuum chamber.

In the process of the experiment the angular velocity variations have been measured with a loop oscillograph, as a function of time, on a signal from the potentiometer. An angle of deflection from the radial direction has been measured visually from the tethered weight deflection from a light wood batten located radially and attached to the rotating part of the deployment mechanism. Via the loop oscillograph the system deployment velocity has been also recorded.

The experimental data on the tethered weight system versus results of the numeric calculation by the technique given in Chapter 3 are shown in Fig. 7.6.

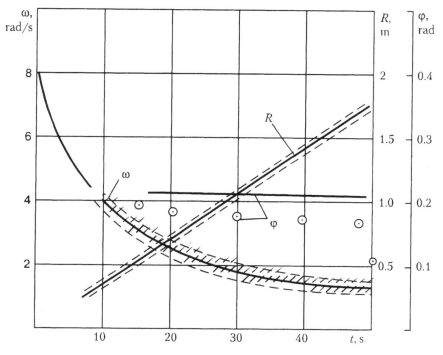

Fig. 7.6. Tethered weight system deployment. ▨ — loop oscillograph;   ⊙ — visual measurements; ———— — theoretical calculation

During the experiments for the solid film sheet re-orientation in space conducted in the vacuum chamber the film sheet of radius $R_k = 2.5$ m with the rigid insert of radius $R_0 = 0.03$ m was spun at a velocity of $\omega = 1000$ rev/min and rotated at an angular velocity of $\Omega = 4-7$ deg/s. There, an amplitude of the reflector edge oscillation run at 20—30 cm and the reverse-travelling period was 60 s. The experimental results are given in Fig. 7.7 where a solid line shows the dependence (7.1).

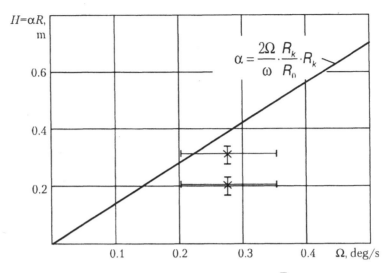

Fig. 7.7. Maximum amplitude of the sheet edge deflection.  ⚹ — visual measurements

Let us proceed to the calculation technique applicability and constraints. The technique has been developed not as a comprehensive theory describing the motion of flexible tether-film systems in space and embracing a vast number of every possible dynamic modes, but as a relatively simple and physically clear mathematical tool to provide the most reasonable dynamic modes of deployment in solving the practical engineering tasks. There, the reliability of deployment by centrifugal forces was given a priority. It should be noted that the new trend of space technology, the structures deployed by centrifugal forces, is under development. This term is not formal, it defines the principle, features, and advantages of the aforesaid structures. The field of centrifugal forces has a dominant role and prevents the system from any move out of the plane of rotation and deviation from the radial direction. The stabilising effect of centrifugal forces must be employed to facilitate the deployment of film structures. The deployment modes have to be organised so that centrifugal forces dominate over other dynamic forces (equations 3.5, 3.6).

The aforesaid feature allows to adopt a simplified mathematical model containing the following assumptions:
  • rectiliniarity of the sheet folds,
  • a lack of oscillations between the film sectors within the plane of rotation,
  • to make calculations, the total mass of all sectors or tethered weights is considered,
  • the actual shape of folds and their shape variation as a function of time are not considered.

The truth of the model has been sufficiently proved by the experiments that provides reasons for applying the calculation technique to structures of fairly larger dimensions.

# CHAPTER VIII

# DESIGN STUDIES ON LARGE SPACE STRUCTURES

## 8.1. Solar Sailing Vehicle Design

The solar sailing vehicle has been designed in 1990 using the experience gained from research accomplished at RSC Energia by then. Works on the solar sailing vehicle design have been initiated by the competition announced in the United States in honour of the 500-year anniversary of the America discovery. The reflector was made of two counter-rotating circular film sheets of separate sectors combined around the periphery. The system assumed the zero angular momentum and was controlled by the deflection of the axis combining two rotating reflectors and by the resulting precession. The design principles inherent in the system have been patented later on. The design was based on the geliorotor concept [1] offering the rotating system deployed by centrifugal forces. The system was made of radially oriented separate strips 8000 m long and 4 m wide coming from one centre.

A main drawback of this geliorotor was the "floppiness" of separate strips that made impossible the implementation of the system inherent principle of control through changing an angle of attack of a blade-strip and thereby offsetting the centre of mass relative to the centre of pressure created by sunlight. The second drawback of the geliorotor was a necessary initial spinning of the system to start its deployment that assumed the use of plasma thrusters and required a stock of working medium to be provided on board.

The aforementioned drawbacks inherent in the geliorotor system have been eliminated in the solar sailing vehicle design through combining separate sectors around the periphery and thus making the sail structure built by centrifugal forces rigid. There, the principle of control of the blade angle of attack has been preserved and implemented in a simple and ingenious way, i. e. the gyroscope precession principle have been employed. For this purpose a counter-rotating, steering sail of outer diameter $D$ = 50 m and inner diameter $D$ = 20 m was incorporated in the structure. An outer diameter of the primary sail was 200 m, an inner diameter — 60 m. The steering sail provided for changing the angle of attack of separate sectors-blades to add to the spacecraft control capability and compensate moments accumulated from the exterior effects. The primary and steering sails were of a similar design, each made of 12 separate sectors of Kapton film and combined around the periphery. To attach the inner contour of the both sails the guying-type fasteners were used with the spools accommodated on the central cylinder. Fig. 1.2 shows the design using a tethered flywheel instead of a steering sail.

The primary and steering sails, both were provided with two deployment mechanisms of a similar design consisting of a central cylinder accommodating spools on which the solar sails were stowed into a transport configuration, the spools being set in rotation by electric drives based on the converter-fed motors of 2 W a.c. power. From the variants calculation using the technique described in par. 3.2 the time required to expand the sails has been estimated as 5.5 h.

The competition assumed the total mass of the entire system of up to 500 kg. Besides the sails, the system incorporated all standard subsystems of unmanned space vehicles. The system has been designed for a mission to Mars. A complete set of technical documentation has been released. An explanatory note describing the preliminary design was covered by 21 volumes. This project headed the list of designs brought up for the competition by the USA, Japan, and France, however, was not implemented. The Znamya-2 deployment mechanism and the reflector structure served as the solar sailing vehicle model scaled to 1:10.

## 8.2. Framed Reflector Design for Illumination of Arctic Regions

In similar designs known to date the large film reflector surface is built using one of the following methods: by centrifugal forces (structureless systems) and by attaching the film to a supporting structure (a frame) to hold it rigid. The concepts thus far verified represent designs of solar sailing vehicle dedicated for missions to Mars and presented by the USA, Japan, France, and the former USSR in honour of the 500-year anniversary of the America discovery by Columbus. The former USSR was presented by NPO Energia. The Energia design was a structure built by centrifugal forces. The rest of the designs were based on framed structures. The primary feature of the Energia design is that the surface accuracy (its reflectance) is imposed more stringent requirements against those placed upon the solar sailing vehicle design. An angle within which an observer from the Earth is viewing the Sun is 30′ and, thus, the angular accuracy to be met during the reflector surface fabrication and its orientation should be of a lower value (10′ is selected) to assure a maximum concentration of light energy in a spot retro-reflected to the Earth surface.

Presently, it is impossible to achieve the similar surface accuracy in structureless systems built by centrifugal forces and offering a number of specific features including the uniquely low mass-dimensional characteristics and simplicity. Among the drawbacks of such structures is the problem of controlling (damping) a wide spectrum of surface oscillations caused by various dynamic impacts (resulting from the deployment from a stowed configuration, re-orientation) which is not resolved thus far. For this reason, in the given project various designs of framed structures different in configuration and location of a rigid supporting structure have been considered. Each of the designs has been preliminary verified and evaluated for strength and mass-dimensional characteristics, as well as for attitude control capability.

The input parameters were such as reflecting surface area, $S = 4 \cdot 10^4 \text{ m}^2$, and film tension force, $q = 35$ N/m.

The following designs of reflector supporting structures have been considered:
• a rigid frame of a triangle form around the structure perimeter,
• a rigid frame made as three beams coming from one centre,
• a circular frame located around the reflector periphery,
• a rigid frame made as 6 beams coming from one centre.

A consideration has been also given to a hybrid design. The design is based on a three-dimensional rigid frame which corners contain a large number of intercompensating, in pairs, small circular reflectors built by centrifugal forces. The essential feature of the last design is its modularity making easy fabrication verification of one module, no requirement for the platform orientation within a working zone (the platform is oriented in the LVLH coordinate system, each of the reflectors being individually beacon-pointed from the Earth to initiate operation) and, consequently, lower stiffness requirements for the entire truss structure (tolerated to be disoriented for several degrees).

The first design has been rejected because three struts were two times as long around the periphery as the total length of struts in the second design and also owing to a large value of moments of inertia relative to three primary control axes resulting from the truss structure mass distribution around the periphery. The third design has been rejected for the same reason.

The second design has been denied from the considerations given thereafter. To obtain a required area of film the truss length has to be very large, $h = 200$ m, with the film deflection of 0.05, the truss overall height being 20 m in the base. The stowage capabilities of this truss are not consistent with the vehicle dimensions.

The six-beam and hybrid designs had much the same sufficient characteristics for mass and layout capability. However, the six-beam design has been accepted to be a base option as being more traditional and less bulky against the hybrid design. Besides, the main drawback of the hybrid design (uncontrollable oscillations of surface expanded by centrifugal forces) was retained and required a solution. At the stage of preliminary studies to point simultaneously a large number of small reflectors having one insufficiently rigid supporting structures to one point on the Earth appeared to be problematic.

A structure had been chosen which consisted of expandable beams made of hinge-combined carbon plastic tubes provided with compression members to deploy the structure. This choice has been made from the comparative analysis against a solid, metal truss having a larger mass sufficient to provide a required stiffness. (Fig. 8.1).

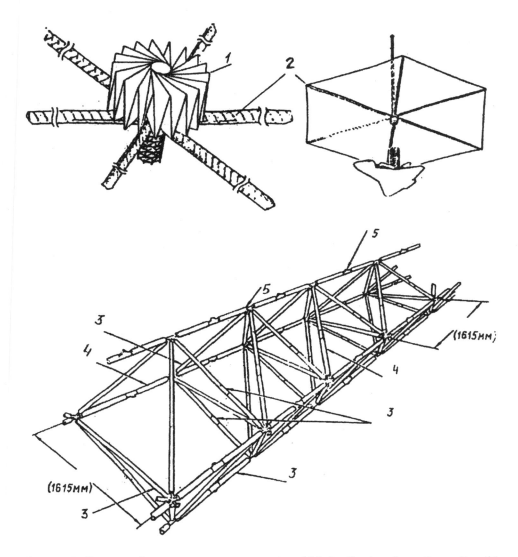

Fig. 8.1. Reflector with a supporting structure. 1 — folded reflecting sheet; 2 — collapsible rigid truss; 3, 4, 5 — truss elemnt

The study results on how to control attitude of a space vehicle of such a size by means of traditional gyroscopic capabilities showed that this task would be intractable. This would require gyroscopes six times as large as those currently available. To control the vehicle through plasma thrusters would require a working medium storage and supply system, $m = 2-3t$, a power supply system, and the vehicle thruster to operate within 10 years. The similar task has been addressed in the Universal Space

Platform Project. The ground verification of the Universal Space Platform control system alone was estimated as $27m.

To control the framed reflector under consideration the gyroscopic principle is employed when the entire reflector structure is spun around the central axis. The generated angular momentum would be compensated via a small flywheel counter-rotating relative to the primary structure. The attitude control is achieved through the deflection of the axis combining the reflector and the flywheel and the resulting rotation of the entire system.

Because the structure rotation is used only to control the system attitude other than to build the surface by centrifugal forces, the required angular rotation velocities and angular momenta will be small.

To fabricate a flat, film reflector it is proposed to use the experience gained by the DADO from designing the solar sail and the reflector for the Znamya-2 experiment.

The flat, film reflector is designed as a solid, film hexahedron which side is 120 m. The reflector material is the polyimide film 8 μ m thick, coated with deposited natrium, the coating being recovered in the course of operation.

For packaging the DADO folding pattern is used (Fig. 3.4).

The flat, film reflector is deployed into the operational configuration simultaneously at the expense of the rigidity of the truss compression members and centrifugal forces generated from spinning the cylinder, being a part of the vehicle, relative to its longitudinal axis and, concurrently with the cylinder spinning, by synchronous unreeling the tethers controlling the deployment process that enables more smooth and sequenced deployment of the flat film reflector a specified velocity.

The system deployment in parallel with its spinning offers a number of advantages against a possible procedure of deployment followed by spinning. Centrifugal forces establish the deployment dynamics, reduce reliability requirements for the actuation of compression members over the first half of the truss length measured from the centre.

The frame-reflector system deployment at the expense of the rigidity of the truss compression members alone may be incomplete and asymmetrical. In this case, the dynamics of the system subsequent spinning would be a problem.

The frame-reflector system deployment is accompanied by the deployment and spinning of the counter-rotating tethered flywheel which mass-dimensional characteristics are well below those of other possible options.

When deployed, the flywheel looks like a tether ring joined with the central cylinder via four cross-like tether links hardly fastened to the main ring. When stowed, the flywheel structure is reeled on four spools accommodated on the central cylinder and combined through the deployment drive. The tethered flywheel is deployed in two steps (par. 3.5).

The space vehicle mass including the reflector run to 3800 kg, the expandable truss making the major fraction of the total.

## 8.3. Large Loop Magnetic Antenna Experiment Accommodated on Board the "Progress-M" Transportation and Cargo Vehicle

*8.3.1. Subject of Investigations*

Electromagnetic VLF waves in the ionosphere and the magnetosphere are a subject of investigations already for more than fifty years. As far as sources of low frequency emissions can be interactions of waves and particles, these emissions bear the information on physical processes in the near-Earth plasma. On the other hand, the observable characteristics of VLF signals and natural emissions are formed by properties of the propagation media, therefore VLF waves can serve a tool of research of a structure and dynamic processes in the ionosphere and magnetosphere of the Earth.

Recently, few attempts of making the experiments on the direct excitation of VLF waves in space plasma were undertaken. The first successful experiment on generation of VLF waves with a loop magnetic antenna has been conducted aboard the orbital complex Mir—Progress-28 in 1987. Comparison of the wave amplitudes measured in the ionosphere during this experiment with the results of the theoretical calculations has shown a good agreement with the used theoretical models of the VLF wave radiation by small loop antennas in the magnetoplasma. Such active experiments in space are planned for study of wave—particle and wave—wave interactions in the ionosphere and magnetosphere and for the investigation of the propagation properties of VLF waves in these media. A peculiar attention is focused on the problem of reception of the signals from satellite-born VLF transmitter on the Earth's surface.

Active wave experiments investigating the processes of modification of the magnetosphere of the Earth under influence of powerful electromagnetic radiation are carried out mainly with ground-based HF and VLF transmitters. It has been shown, that the external electromagnetic wave fields injected into the near Earth plasma from these transmitters affect the characteristics of radiowave propagation, initiate various geophysical and biological processes.

Dealing with the orbital transmitter we have the source directly in the magnetoplasma where it is possible to radiate waves with a wide variety of vector directions, including perpendicular propagating waves. Different ionosphere regions along the orbit are also available for study of the radiation and propagation properties under different conditions.

The Parameter experiment uses the large cable loop antenna (CLA) of 300 m in diameter (Fig. 8.2) as a VLF emitter on board the vehicle. The radiated electromagnetic signals and stimulated emissions are planned to be measured on board the subsatellites. Besides, CLA is proposed to be employed as a tool for parametric generation of VLF waves in the ionospheric plasma.

Large space structures such as metallic film reflectors, cable systems, etc. have been developed at the RSC Energia in early 1980s. The possibility of deploying large, dynamic, tethered structures was demonstrated during the experiment on February 4, 1993, when the circular system containing the film reflector of 20 m in

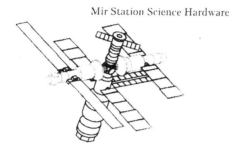

Mir Station Science Hardware

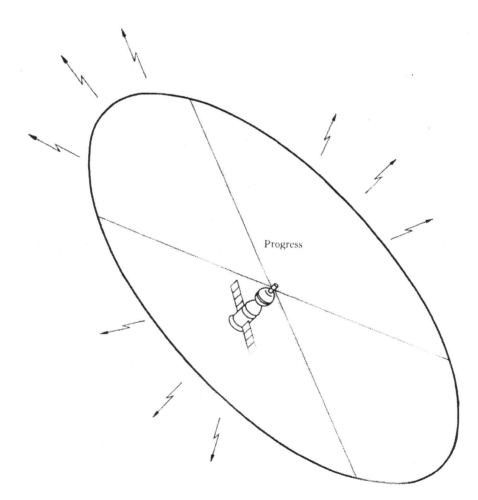

Progress

Fig. 8.2. Cable loop radio antenna $D$=300 m deployed in space by centrifugal force to study radiation and interaction with plasma of strong ELF/VLF waves

diameter has been deployed in space aboard the Progress-M vehicle. Developmental works on such system as the 300 m large, cable loop antenna shaped in orbit by centrifugal forces is proposed to continue with the Parameter experiment.

### 8.3.2. Objectives and Expected Results of The Parameter Experiment

The main purpose of the Parameter experiment using a large CLA is to investigate the processes of VLF wave generation, propagation, and interaction effects taking place in the ionosphere and the magnetosphere of the Earth. Scientific program of the experiments includes:
  • investigation of the large loop antenna and the parametrical antenna system efficiency in generating the ELF/VLF electromagnetic waves in the ionosphere,
  • research in the processes of propagation and transformation of very low frequency electromagnetic waves in the ionosphere and the magnetosphere including non-linear modulation of amplitude and spectrum of waves, generation of trigger emissions, non-linear focusing and ducting, etc.,
  • investigation of geophysical effects of powerful VLF radiation such as heating and acceleration of ions and electrons in the ionosphere, stimulated precipitation of particles from the Van Allen belts, excitation of atmospheric emissions and artificial stimulation of ELF/VLF waves.
  The following results of the experiment are expected:
  • improved design and deployment techniques for the multi-sector loop antennas of a large diameter built by centrifugal forces,
  • the data to be acquired from these experiments will enable to quantify the radiation characteristics of such antennas in the VLF frequency band using the measurements in the ionosphere and on the surface of the Earth.
  To implement this scientific program it is necessary to accommodate the large 300 m circular antenna on the Progress-M vehicle and to provide magnetic moment of the radiator not less than $10^6$ Am$^2$. Measurements of the radiated VLF and ELF waves should be carried out in the ionosphere and on the ground. The input impedances of the antenna sectors and the current intensities should be also measured in the process of the experiments.
  The cable loop antenna built by centrifugal forces mentioned above as the most rational design could be deployed only in orbit. The deployment would be impossible under the ground conditions.

### 8.3.3. Principle Characteristics of the Dynamic Cable Loop Antenna

To implement the experimental system (Parameter experiment) and to make the radiation of VLF electromagnetic waves in the ionosphere effective, the quasicircular cable loop antenna of 300 m in diameter must be deployed. The antenna must be operated at the constant operating frequencies with a current intensity of 15−30 A.

Geometrically and electrically the circular antenna contains 4 sectors (Fig. 8.2) which should be powered coherently to provide the identical direction of a current in the periphery conductor.

The process of the antenna shaping is shown on Fig. 1.3. The central cylinder containing the reeled up antenna rotates around the axis normal to the antenna plane. Then the circular and radial cables are released in turn. Further, the CLA configuration is maintained by centrifugal forces, since the antenna continues to rotate with constant angular speed.

Principal parameters of CLA:
- diameter: 300 m,
- diameter of the conducting cable: 2.2−2.5 mm,
- thickness of polymer insulation: 1.0 mm,
- operating current intensity in the antenna: up to 32 A,
- angular velocities of rotation: 9.9 rad/s (initial) and 0.1 rad/s (final),
duration of the antenna deployment: 5−6 h.

To monitor and control the antenna configuration, it is supposed to place visual elements (such as "tags" of a thin reflecting film) on the cable circular contour of the antenna.

To determine the optimum modes of the antenna deployment, the deployment process has been mathematically simulated. The obtained results confirmed that the in-orbit deployment process of the 300 m cable loop antenna could be stabilised. The mathematical simulation of the antenna as a dynamic mechanical system considered the influence of the carrier in-orbit motion.

Mechanical elements of the CLA deployment mechanism have been tested under the appropriate conditions in the field of gravitational forces. On the basis of the laboratory investigations, the designs stable during the deployment steps have been defined.

### 8.3.4. Progress-M Vehicle Design Features Required for Parameter Experiment Implementation

Orbit parameters:
- inclination: 51.6°,
- altitude: 350−450 km.
Time for operations:
- general: up to 33 days;
- in autonomous flight: up to 30 days;
- for the experiment implementation: up to 27 days.

The CLA experiment aboard the Progress-M transportation and cargo vehicle would be possible at the end of its re-supply mission (delivery of equipment, liquid propellants etc. to the orbital station).

The accommodation and installation of the scientific devices aboard the vehicle is planned in way to minimise the construction complexity. The antenna deployment mechanism in a stowed configuration, the VLF generator and other equipment would be accommodated in the cargo compartment. Six independent power supply units (accumulator batteries) will be delivered outside the pressurised cargo compartment.

Upon completion of the re-supply mission, the antenna deployment mechanism will be prepared for service on the vehicle hatch. Two TV monitors will be provided on the vehicle to control the cable antenna deployment process.

One radiation period will be up to 10 min. With the restricted power consumption the radiation period could be decreased, or three subperiods 2—3 min each would be observed.

Two to four radiation periods per day are assumed. The Parameter experiment duration and radiated power depend on power resources of the vehicle of the Progress-M type. However, the experiment duration should not be less than two weeks.

The control and measuring complex should provide an opportunity to employ the Progress-M vehicle and subsatellites within the coverage of the ground telemetry stations. This opportunity could be provided in the magnetically conjugated regions of northern and southern hemispheres by the logistic vehicles, or by the ground based stations.

The ground support program of the experiments includes the registration of the ELF/VLF electromagnetic waves radiated by CLA and the stimulated emissions, Doppler and amplitude—phase measurements, radio beacon and optical observations for monitoring the ionosphere perturbations initiated by the VLF waves.

## 8.4. Space Debris and Particulate Matter Registration and Collection System

### 8.4.1. Problems of Space Debris

It is suggested to develop an orbital system enabling to record, collect, identify, and analyse a composition of space debris and particulate matter of both technogenic (orbital debris) and natural (micrometeorites) origin and varying in size from 0.01 to 10 mm. To obtain adequate information the following methods are proposed:
- contact methods using large-area film detectors (from 500 to 1500 $m^2$),
- optical remote (passive and active) methods.

The existing methods of recording the space debris and particulate matter are inadequately efficient to make the proper assessment of spacecraft damage probability. Ground observations of meteorites deal with the particles entering the atmosphere at cosmic velocities and would miss the particles captured up on artificial satellite orbits. These observation are spatially limited. Radar methods make it possible to record separate particles of at least 5 cm in size. Finally, the detectors of meteorite sensors feature a small area. For this reason they are hardly suitable for analysing the statistical distribution of particles ranging beyond 10 μ in size. The particles varying from 0.1 to 10mm pose the maximum and increasing in time hazard to spacecraft due to their great kinetic energy.

The key problem of how to assess the probability of spacecraft damage by space particles is attributed to their small concentration which decreases dramatically with the particles size increase. Thus, to obtain the reliable statistics a set of long-term observations is required.

Since the information on the concentration of debris and particulate matter in the orbit of 400—450 km in altitude is rather urgent, primarily, to assure safety for the International Space Station Alpha under development, it would be desirable to design a high-performance detector that could be placed in orbit within the forthcoming 2—3 years. Therefore, the main emphasis at the first phase of activities should be placed upon the development of the detector which would be simple in design and easy to implement. At the second phase of activities more sophisticated systems will be considered.

To efficiently record debris and particulate matter of the above sizes, large surfaces of detectors are required whose area, with the traditional design solution, is limited by the existing launching capabilities. The suggested design solution is based on the RSC Energia experience gained from designing large film structures expanded by centrifugal forces (solar sails and reflectors) and from designing hardware under international projects.

### 8.4.2. Recoverable Detector

Fig. 8.3 shows the detector consisting of strips of polymer film or metal foil which are radially expanded by centrifugal forces when prepared for operation. In the transport configuration the strips are reeled up on a drum. Centrifugal forces required to unfold the strips are generated by the detector deployment mechanism through its spinning. The deployment mechanism consists of a hollow cylinder casing incorporating a built-in drum and electric drives. The drum is designed like a "squirrel-cage". Ends of the strips reeled up on the drum are released between fillets located around the drum periphery and joined together by means of filaments. The drum is accommodated on a fixed axis secured to the Progress-M vehicle docking unit. The deployment mechanism casing is kinematically linked to the axis through electric drive (1), the drum being is kinematically linked to the casing through the other electric drive (2). Drive 1 rotates the deployment mechanism at an angular velocity generating centrifugal forces sufficient to release the strips. Drive 2 rotates the drum relative to the casing thus making the strips to release at the velocity required for a stable unfolding.

To unfold the strips, drive 2 is set in the reverse rotation at which the deployment mechanism slowdown is provided by drive 1. The deployment mechanism is installed outside the Progress-M cargo compartment, on the docking unit. A reaction moment generated as the strips are being unfolded and folded is compensated by the vehicle attitude control thrusters. The aerodynamic drag is compensated by the vehicle approach thruster. With the detector area of about 1000 m$^2$, the Progress pro-

8.3. Deployment from stowed configuration

pellant stock would be sufficient to expose the detector in the orbit of 450 km within a without its lowering.

Upon completion of the experiment the strips are folded into the transport configuration. The deployment mechanism together with the folded strips are jettisoned from the Progress cargo vehicle. A braking pulse is sent to the Progress vehicle for its de-orbiting and subsequent destruction. The deployment mechanism will be recovered on Shuttle. Upon the recovery, the detector deployment mechanism will be explored for a number and size of holes. Besides, The dispersed particulate matter will be analysed. Photometers recording the passage of particles could be installed on the Progress vehicle to obtain rapid information, as well as TV cameras with a wide field of vision to obtain the detector images.

### 8.4.3. Unrecoverable Detector

This detector differs radically from the recoverable detector in using a film disc instead of strips, the disc consisting of separate sectors combined around the periphery and reeled up on spools to be deployed in space. A similar structure has been tested in the space-based Znamya-2 experiment.

The disc structure of the detector differs favourably with the strip structure in that, having similar linear dimensions, it offers a sufficiently larger area. However, the disc structure, despite the possibility to be transported in a small volume package, gives no way for the film to be automatically unfolded and re-packaged for retrieval. Thus, it appears necessary to record and analyse the detected particulate matter in real time. This would require to develop special techniques and instruments.

Measurements will be performed in a way described thereafter. Sectors of film discs are electrically insulated from each other. Electric voltage is applied to the discs. Thus, the detector will serve as a condenser. Parameters and properties of dust particles will be recorded from the outcomes of the film penetration by the particles, formation of the plasma luminous discharge accompanied with an acoustic effect in the film. Films of materials with the space-charge polarisation could be used, for example, polyvinyl-fluoride, which penetration is accompanied with a short current pulse in the exterior circuit.

To measure the particulate matter parameters employing the optical and acoustical equipment, the following instrumentation is considered:

• an optical-electronic instrument for recording the flare integral light energy,

• multi-channel photometer for measuring the volumetric coefficient of particulate matter dissipation,

• an acoustic meter of film oscillations,

• a spectrometer for analysing products of flares resulting from the film penetrations.

In developing the instrumentation it is proposed to employ the experience gained from the Venus-Halley International Project (using such instruments as Photon, Dusma, SP-2 in the capacity of analogues).

To generate the impact pulse required to record particles when they hit the film material, the following mechanisms are considered:

• processes of pointing a charge and current in the film condensation,

• a flare attending the penetration of the first film layer.

The kinetic energy of a particle can be judged from the acoustic pulse energy. With the distance between the disc sectors known, a particle velocity is determined from the time difference between pulses. From coordinates of the penetration points the particulate matter trajectory is found. A chemical composition of fast particles creating the plasma cloud when the film is penetrated can be determined by the spectral techniques.

A key issue in the development of the space debris and particulate matter recording system using a large area film surface is how to optimise the deployment and re-packaging control of the transformable structure elements. The task is to establish laws of control for motion or braking moments and deployment or repackaging velocities to prevent the system from the intolerable oscillations and getting in a tangle, the control being performed in the required time frame.

The most suitable for deployment and repackaging is the surface structure composed of separate strips of metal foil (aluminium, titanium).

The combined Equations describing the deployment dynamics of the given structure deployed from the central cylinder under the control law is presented in Chapter 3.

## CHAPTER IX

# FEATURES OF DESIGNING APPLICATION SPECIFIC, LARGE STRUCTURES. GENERAL PROBLEMS

## 9.1. Basic Design Criteria and Parameters

The principal steps of designing a space vehicle based on the deployable, large structure has a number of features associated with the nature of the resulting product.

Just as in designing other types of space vehicles, the given design process is started from analysing the application, establishing the product efficiency criteria and requirements emerging from the application.

Let us consider the task of creating the solar sailing vehicle. Similar to a transportation vehicle, a solar sail has to generate a certain characteristic acceleration (of the order of $a = 10^{-3}$ m/s$^2$) thereby limiting the acceptable specific mass of the vehicle, including the sail and service systems, of the order of $\rho = 5 \cdot 10^{-3}$ kg/m$^2$. The possibility of lofting the vehicle in orbit would impose requirements for its accommodation on the carrier, the requirements depending on the assembly packaging coefficient, $K$, and dimensions of the largest element. Dimensions of the solar sailing vehicle being designed are limited by the dimensions of the carrier cargo bay at the top and by the possible minimum sufficient for the mission implementation at the bottom.

The sail-produced thrust depends on the specular surface quality, i. e. with the reflective coating, twice as large pulse will be intercepted by the reflector surface from the sunlight quanta against the black coating. The light reflection can be both mirror and diffuse.

The mission duration imposes requirements for providing strength and reflectivity of sail material exposed to space environment and operation effects.

The orbit are required to be out of the Van Allen belts (see par. 5.5).

All systems of the solar sailing vehicle are imposed requirements for the enhanced reliability likewise a free-flying unmanned spacecraft designed for a long-term service life. The reliability is assured through the proper selection of design solutions and the product integrated reliability assurance test plan.

In designing the application specific space vehicles, such an important criterion as a cost for one kilogram of payload placed in orbit, is taken as a cost for one kilogram of scientific hardware to be used for obtaining scientific data, or solving any specific task.

Dynamic characteristics of processes of deployment from a stowed configuration are described by parameters $\omega^2/\dot{\omega}$ and $\omega R 2 v$ (see par. 3). Re-orientation in space is defined by parameters $\Omega/\omega$ and $R_k/R_o$ (see par. 4).

The perturbations caused by the gravitational field gradient are assessed by parameter $\omega_{orb}/\omega$, where $\omega_{orb}$ is the orbital rotation velocity. Stresses occurring in the rotating structure are of the order of $\sigma = \rho \omega^2 R^2$, strength margin $n = \sigma/\sigma_{accept}$, where $\sigma_{accept}$ is the acceptable stress evaluated during tests for long-term strength and yield considering the combined impact of orbital factors.

The product service life is taken as the most important parameter governing the configuration, the selection of structural materials, the scope of ground tests, and, finally, the product cost.

Thus, in designing a solar sailing vehicle, the characteristic acceleration, $\alpha$, is a key factor directing all design goals such as the desire to reduce the sail specific mass $\rho$, to enhance the surface reflectivity, while leaving the possibility for the task to be solved even with the black coating and, consequently, with no high requirements for the reflecting surface performance and attitude.

The sail canvas can be designed both solid and split. Because of a small value of tangential stresses, a split sheet would be less reflective against a solid one.

Similar requirements are placed upon the reflector sheet structure in the task of creating a passive retransmitter for the TV, radio, and laser communications, false targets and protective shields, as well as a shield for removing space debris. In all tasks listed above no strict requirements are placed upon the surface accuracy, reflectivity, and attitude.

On the contrary, in designing a reflector of a system intended for illuminating the Earth regions with reflected sunlight the specular surface accuracy and the mirror pointing would be given a top priority. In the given case the intensity of illumination of a specified region is a key parameter depending on the reflector area, the surface reflection factor, absorbing and scattering properties of atmosphere, the angle of light incidence to the Earth surface. It would be desirable to have the reflection factor the maximum close to one over the entire surface and the accuracy of the beam pointing from the orbit must be well beyond the angle of 32′ at which the Sun could be seen from the Earth.

The aforesaid parameters are deciding in designing the reflector and defining its structure and pointing system complicity. An important feature is that the reflector would be required to be permanently re-oriented at the preset accuracy of pointing to the specified area of the Earth while moving in orbit that would make the control system more complicated. To achieve a high reflection factor, the reflector canvas structure has to be maintained in the two-dimensional stress state close to the elastic strain limit along both axes. This would be feasible with the reflector using the supporting structure (see par. 8.2), the system where the vehicle is of a low specific mass, or with the reflector having no the supporting structure provided that special measures are taken to produce a uniform, two-dimensional stress over the most part of the reflector surface with oscillations from gyroscopic forces actively damped. It should be noted that the task of illumination is the most complicated among the feasible applied tasks.

A number of important tasks exists with no requirements for the structure re-orientation. For example, the use of a rotating, structureless, parabolic, film concentrator for replenishing the ozone layer around the Earth. The evaluation results indicate that to solve the task, 30 Sun-pointed concentrators of 1 km in diameter would be required on the solar synchronous orbit of altitude H = 1680 km. In the focus of the concentrator a power unit steam generator has to be accommodated to feed a powerful laser operating at the wavelength exciting the energetic levels of oxygen in the upper atmosphere and contributing to the production of ozone under the solar radiation. The concept of interest is the direct illumination at a tangent to the Earth of the ozone layer with the film reflector plane in the solar synchronous orbit.

No re-orientation of the tethered ELV/VLF antenna is required. The antenna plane is located within the orbit plane and, in solving the application specific task, the antenna would behave as it were "rolling" along the orbit.

Extremely attractive is the employment of structureless, rotating systems in the future lunar technologies. The Moon has no atmosphere and its gravitation is 6 times less as compared to the Earth. It thus appears that the Moon is a unique celestial body with respect to the use of rotating, structureless systems. The architecture of power generation facilities should be based on precisely such structures as mentioned above. Surfaces of solar arrays, solar concentrators of power units, high power transmitting and receiving antennas could be used for this purpose. The gravitation effect on the Moon is evaluated from criterion $\omega^2 R/q$, where q is the gravitational acceleration on the Moon. From the current predictions, the Moon holds the great promise for supplying energy to the Earth owing to the sufficiently larger stocks of fission fuel. The stepwise exploration of the Moon is being projected including the transfer to the Moon of ecologically harmful and power-consuming technologies from the Earth. The structureless, large, rotating systems are undoubtedly contributing much to the reduction of capital investments.

For the majority of tasks under consideration it would be reasonable to employ deployment mechanisms based of the electromechanical type based on electric drives offering high reliability and suitability for ground verification. However, to deploy protective shields, a gas jet would be preferable because of no need in the deployed shield re-orientation, the lack of the counter-rotating structure, and the justification for more complicated ground verification in the event of serial production. The gas jet is capable of providing a high velocity of deployment. The deployment mechanisms can be accommodated on the multi-spool and single-spool stowed patterns of film canvas. The single-spool design would enable to achieve a high packaging coefficient.

To re-orient a large, rotating structure, the principle of gyroscopic control is employed when the structure itself is serving in the capacity of a gyroscope. Through introducing the rotating film or tether structure the sum angular momentum of the system is brought to zero and the system is re-oriented at the expense of deflection of the axis combining the counter-rotating structures and the resulting precession. The axis deflection occurs in a hinge with two degrees of freedom which is controlled by precessional drives. The control goals require the axis to be deflected at a very small angle ($\approx 1°$) to a high accuracy as per the required law. The drive operates in a reverse mode and provides the "point-to-point" deflection within the control angle. There is every prospect of using drives based on piezoelectric crystals and offering a large moment at a small stroke, precessional control, and a lack of dead stroke which would be inevitable in gear transmissions.

## 9.2. Perturbations and Limitations Associated With In-Orbit Environment Effects

Let us consider how a large structure is effected by different orbital factors such as: residual atmosphere at various attitudes, gravitational field gradient of the Earth and other celestial bodies, radiation fluxes from the Sun and Galactic cosmic ray flux, the Van Allen belts, the Earth's magnetic field, etc. The factors listed above would impact the vehicle performance characteristics such as its service life, capabilities for re-orientation in space, and surface accuracy.

A content of residual atmosphere at different attitudes is varying by an order and a half depending on a large number of factors: season, solar activity, geographical latitude, etc. The residual atmosphere would created the aerodynamic drag to the vehicle and define a corrosive interaction with the structure elements, as well as the destruction of film materials.

Table 8 shows preliminary data on the total density of residual atmosphere (for all components) at altitudes ranging from 100 to 2500 km.

*Table 8*

| $H$, km | 100 | 200 | 300 | 400 | 1000 | 2000 | 2500 |
|---------|-----|-----|-----|-----|------|------|------|
| $\rho$, kg/m³ | $5 \cdot 10^{-7}$ | $8 \cdot 10^{-11}$ | $2 \cdot 10^{-12}$ | $8 \cdot 10^{-14}$ | $5 \cdot 10^{-15}$ | $1 \cdot 10^{-15}$ | $5 \cdot 10^{-16}$ |

At attitudes over 300 km a path length of particles is well beyond the possible dimensions of the vehicle with the flow being free-molecule. In the given case a drag force can be evaluated by the Equation

$$X_n = C_x q S,$$

where $C_x$ — a drag coefficient depending on the accommodation of particles on different surfaces ($C_x \approx 2$);

$q = \rho\, V^2 / 2$ — a dynamic head;

$S$ — an area of surface normal to the flux.

In performing the precise calculations for complicated-shape bodies, $C_x$ is determined experimentally through purging a vacuum chamber and thus simulating a real flux and the body configuration.

For the reflector operating in orbit, its drag could be evaluated assuming that the reflector is of a flat configuration with S being the reflector area, $V$ — orbital velocity ($V \approx 8$ km/s), $\rho$ — density as per Table 8, $C_x = 2$.

With $S = 10^4$ m$^2$ and $H = 350$ km (the altitude of Space Station Mir) $X_n = 0.2$ N; with the altitude of 2000 m (below the Earth's radiation belts) $Xn = 6 \cdot 10^{-7}$ N, i. e. negligibly small. With $S = 10^6$ m$^2$ and $H = 350$ km the drug is of a sufficient value, $X_n = 20$ N. For the altitude of 2000 m $X_n = 6\ 10^{-2}$ N, i. e. remains as small as before.

A drug of a plate in a lateral free-molecule flow can be evaluated from the expression

$$X_\tau = \sigma / 2\,\pi^{1/2}\,\rho\,C_m\,C\,S,$$

where   $\sigma$ — the diffuse reflection factor;

$C_m$ — the most probable velocity of gas particles ($C_m = (2\,k\,T/m)^{1/2}$);

$C$ — the surface motion velocity.

The $X_\tau$ is evaluated as one order of magnitude lower than $X_n$.

A moment of friction acting on the rotating surface on its both sides is derived from the equation

$$M = 0{,}5\,\pi^{1/2}\,\sigma\,\rho\,C_m\,\omega\,R_k^{4},$$

where $R_k$ — the system radius;

$\omega$ — the circular velocity of rotation.

With $H = 350$ km, $R_k = 100$ m, $\omega = 0.1$ rad/s, $M = 10^{-3}$ N m that is much below the friction in bearings.

In low orbits, from 300 to 400 km, residual oxygen would cause all types of exposed films to be intensively destructed thereby reducing the service life of the film structure by the order of several dozens of hours depending on how the film surface is oriented to the incoming flow (see par. 5). At altitudes of $H > 1000$ km no interaction with oxygen is observed and the film would be destructed mainly due to the UV solar radiation. Within a range of altitudes from 2000 km to 6000 km the Van Allen belts would contribute to the destruction.

A braking moment resulting from the effect of the Earth's magnetic field on the rotating, conductive disc is obtained from the equation considering the insulation of the generated electromagnetic field by ambient plasma:

$$M = 2\,(n_i\,B)^{1/2}\,/(e\,C)\,m_i\,V\,\omega\,R_k^{3},$$

where $n_i$ — plasma concentration;

   $m_i$ — average mass of ion;

   $B$ — strength of the Earth's magnetic field;

   $e$ — electron charge;

   $C$ — light speed.

A value of this moment is four orders below the moment of friction.

In orbit an expanded structure tends to expand under the gravitational field gradient and to occupy a position in the direction of the Earth centre.

With the disc oriented at angle $\alpha$ to the orbital velocity vector a value of moment relative to the axis coming through the disc centre is calculated from the Equation:

$$M_q = \pi / 16\, q\, \mu\, R_k^4 / R_3 \sin 2\,\alpha\,,$$

where   $q = q_3 (R_3 / R_0)^2$ — acceleration of gravity in the orbit of radius $R_0$;

   $R_3$ — radius of the Earth;

   $q_3 = 9.8$ m/s$^2$;

   $\mu$ — surface density of the disc, kg/m$^2$.

With $\alpha = 45°$ $\mu = 5\ 10^{-3}$ kg/m$^2$, $R_k = 100$ m, $H = 1000$ km, $m_q = 0.2$ N m that is one order of magnitude less than the moment required for re-orientation at an orbital velocity of $\Omega$.

$$M = \mu\, R_k^4\, \omega\, \Omega = 5\ \text{N m for}\ \omega = 10^{-1}\ \text{rad/s and}\ \Omega = 10^{-4}\text{rad/s}.$$

Orbital factors such as lunar and sun gravitation, as well as structural and design factors like the inside friction between the cable fibres and residual stresses have been evaluated as practically ignored in actual engineering structures.

No resonance from longitudinal, transverse, and torsional oscillations can be tolerated because their frequencies are fairly exceeding the frequencies of possible operational effects.

## 9.3. General Concepts and Problems

Based on the foregoing material, let us sum up the basic concepts of designing large space structures expanded by centrifugal forces enabling to:

   • depart from the traditional design, use specific orbital conditions (high vacuum, microgravity) and centrifugal forces for developing the application specific structures,

   • provide a zero angular momentum for the system through introducing the counter-rotating flexible elements enabling the system deployment from a stowed configuration and the non-propulsive re-orientation of the deployed and rotating system,

• re-orient the deployed rotating structures in space using the gyroscopic principle at the expense of precession occurring as the axis combining two counter-rotating elements of the structure is "deflected",

• employ the multi-fold stowed patterns for film and tethered systems to provide stability through the deployment due to their multi-linkage and symmetry,

• use solid film structures incorporating additional design considerations to provide the uniform, two-dimensional, stressed state in developing reflectors,

• deploy the structures from a stowed pattern using the inverse dependence of the moment on the speed of the electric drives based on the mathematical simulation of the deployment dynamics,

• employ polyimide film materials with the restorable natrium coating in systems using the long-term service life reflector,

• actively damp oscillations of the rotating surface as being re-oriented in space to avoid effects of gyroscopic forces, or select a modular system wherein such oscillations would be ignored,

• perform the product integrated reliability tests and the increment ground verification including the mathematical simulation of force acting on the structure elements and the most important steps of testing the electromechanical hardware.

The key issue to cope with in designing film reflectors aims at obtaining a high accuracy of the reflector shape and orientation and a high quality of the reflective surface. The requirements listed above define a scope of primary research activities:

• research in the film material long-term strength and creep and reflective coating degradation,

• development of systems to restore the reflective coating in space at a high accuracy of orientation, reflectance control,

• development of the vehicle attitude control system providing a high accuracy orientation in space (10′),

• development of the system capable to monitor and control the reflector surface shape.

A long-term service life of the product ($T = 10$ years) imposes stringent requirements upon its reliability. Research of materials should be aimed at developing techniques for simplified tests and methods of forecasting when exposed to the combined effect of operational conditions. Additional efforts would be required to study how optical and mechanical properties of metallized polyimide film are changed under the combined effect of solar electromagnetic radiation, protons and electrons using the reliable techniques to simulate operational conditions.

To control a space vehicle at such a high accuracy calls for creating the appropriate control algorithms and developing the hardware for their implementation. A large scope of design-theoretical research would be required to optimise the control tasks.

To solve problems on the surface state monitoring and pointing to a specified area to be illuminated, the application specific hardware would be required.

The principal feature of the flight structure verification is that no way is found to test its primary elements — a film reflector and a tethered counter-rotation flywheel— under ground conditions because of the lack of so large vacuum chambers ($D = 200$ m) and the availability of gravitational forces.

For this reason, all elements of the flight structure responsible for the deployment dynamics of the stowed reflector and flywheel into the operational configuration and providing the controllable rotation of the reflector in space have to be tested including simulation of actual loads applied through the entire set of standard tests.

Stringent requirements for the product reliability call for the development of integrated test set-ups and necessitates a large amount of work to be performed.

The following test set-ups are required:

- a test setup to verify deployment and rotation drives of the reflector and the flywheel is intended to test and verify a driving system including the electromechanical elements and the onboard computer software,

- a test setup to simulate the deployment dynamics should contain systems capable to hang and simulate loads induced by centrifugal forces.

The film reflector tests shall include:

- tests for strength during transportation in a stowed configuration,
- mechanical tests simulating the in-flight load acting on a stowed reflector,
- kinetic tests simulating the environment suited for storing the film sheet.

The system elements and assemblies should be tested for vibration strength and temperature cycling under the standard programs and techniques used in developing the space technology products.

# References

1. Friedman L., Carroll W., Goldstein R., "Solar Sailing — the Concept Made Realistic", AIAA Pap., 1978, N82, pp. 1—16.

2. Patent 2486722 (France, HO1Q 15/20, "Reflecteur D'Antenne Deployable/Nationale Industrielle Aerospatiale, 1980.)

3. V.A. Koshelev, V.M. Melnikov, S. Yu. Zaitsev, O. Yu. Krivolapova, G. Pignolet, "Znamia-2: From Mathematical Deployment Simulation to Prospects for Space Bulky Structures Formed by Centrifugal Forces", 44-th International Astronautic Congress, Paris (France), 1993, pp. 246-248.

4. V.A. Koshelev, V. M. Melnikov, S. Yu. Zaitsev, "Large, Deployable Space Reflectors Formed by Centrifugal Effects — Design, Development, and Dynamics", 45-th Congress of International Astronautics Federation, Jerusalem (Israel), 1994, p. 118.

5. V.M. Melnikov, S. Yu. Zaitsev, O. Yu. Krivolapova, "Dynamics of Space Structures Formed by Centrifugal Forces", International Aerospace Congress (IAS-94), Moscow (Russia), 1994, p. 191.

6. V.A. Koshelev, V. M. Melnikov, A. K. Gorodetsky, "Registration and Collection System for Cosmic Particles", 46-th IAF Congress, Oslo (Norway), October 2—6, 1995.

7. V.A. Koshelev, V. M. Melnikov, V. A. Komkov, "Prospects for Designing Large Space Constructions", 46-th IAF Congress, Oslo, Norway, October 2—6, 1995.

8. V.P. Nikitsky, V. A. Koshelev, V. M. Melnikov, Yu. E. Levitsky, "Space Debris Registration System Using Large Surfaces", 1-st International Workshop on Space debris", Moscow, October 9—11, 1995.

# Index